D0098214

ENGLAND AND FRANCE IN THE MEDITERRANEAN

DA 471
.L7
1970

ENGLAND AND FRANCE IN THE MEDITERRANEAN

1660–1830

BY

WALTER FREWEN LORD

KENNIKAT PRESS
Port Washington, N. Y./London

INDIANA
PURDUE
LIBRARY
OCT 1973
FORT WAYNE

WITHDRAWN

ENGLAND AND FRANCE IN THE MEDITERRANEAN

First published in 1901
Reissued in 1970 by Kennikat Press
Library of Congress Catalog Card No: 73-110912
SBN 8046-0894-6

Manufactured by Taylor Publishing Company   Dallas, Texas

PART I

GIBRALTAR—ALGIERS

ENGLAND AND FRANCE
IN THE MEDITERRANEAN

I

IN all probability England would have played but a small part in the Eastern Question, as it is understood to-day, if it had not been for Napoleon Bonaparte.

It is not to be expected that Continental students of the history of England will accept any such conclusion. Seeing Gibraltar (which is, geographically speaking, Spanish ground) in English hands, they cannot but conclude that the "key of the Mediterranean" was grasped with some intention of inaugurating a policy which appears to have developed so continuously and so portentously in the course of the last two hundred years. But the listless attitude implied by Lord Tyrawley's remark to Henry Fox (August 26, 1756)—"I do not see that we do ourselves much good, or anybody else any hurt, by our being in possession of it"—thoroughly characteristic as it is of all the dealings of England with the Rock of Gibraltar, does not appear to bear out this view. The six attempts to shake off the incubus render such a view untenable.

It is the same with Malta. Seeing England in possession of a fortress which is (geographically speaking) Italian ground, it is impossible to avoid ascribing the anomaly to ruthless acquisitiveness on the part of England. Yet we have but to study the negotiations which preceded the Treaty of Amiens in order to arrive at a different conclusion.

Two great names—those of William Pitt the elder and William Pitt the younger—typify British rapacity to the Continental observer. The elder Pitt did his best to disembarrass England of Gibraltar ; the younger Pitt did his best to avoid acquiring Malta. These are material reflections.

In the face of the audacious annexation of Corsica, the unprincipled proposal to annex Sicily, and the dramatic acquisition of Cyprus, it takes some courage to maintain that measures so startling were not part of a profound policy. Cyprus may pass ; but the annexation of Corsica was an act of bewilderment rather than of policy, and the Sicilian scheme was disavowed promptly, and caused genuine vexation to the British Cabinet.

It is when the result of two hundred years of effort is contemplated, it is when we consider England in Egypt as an important factor in the Eastern Question, that we encounter the most bitter comments, that we hear most of the long and unscrupulous plotting which has led England to that prominent position. But in the face of the agitation, the scares, the hurried and anxious movements of 1798, of 1803, of 1807— of the whole course of England's dealings with Egypt, the charge against England of a Machiavellian policy is a patent absurdity.

In the sense that the Vatican has a " policy," that St. Petersburg, or that Pekin has a " policy," England had no policy at all in these apparently connected and concerted measures. For the purpose of study, it is not unfair to say that " the great magnet of India drew us along the perilous waterway." As a matter of history, the Eastern Question in its present form was produced by the activity of the Emperor Napoleon. To the same source is to be ascribed the change of attitude on the part of England during the years 1795– 1815. To the removal of that influence is to be ascribed the lassitude of the Foreign Office in the years succeeding 1815. To the resumption of the policy indicated by the first

Napoleon is to be ascribed the later activity of France on the African shores of the Mediterranean, as well as that uneasiness on the part of England which has transformed and hardened, under popular agitation, into something resembling a policy.

The complicated web of interests which is known as the Eastern Question had hardly assumed its present shape two hundred and fifty years ago. The thorny problems of the Lebanon, the Danube, the Crimea; of the guardianship of the Holy Places, or the suzerainty of Constantinople over Egypt, had not yet troubled the sleep of European diplomatists. Under the all-embracing mantle of Turkish dominion, those potentialities of distraction for Europe remained unrevealed and unsuspected. The very phrase, the Eastern Question, was not yet invented. In so far as the East troubled the West, the sources of discord were to be found two thousand miles nearer home than Alexandria.

It is rather of the Mediterranean Question that we should speak in considering the relation of England to the East two hundred and forty years ago.

What we are now accustomed to call the East was at that time more remote than any part of the known world is to us at the present moment. Much of it was even undiscovered. Therefore the trade to the Levant (partly because of its greater comparative value than the same trade to-day, partly because of the non-existence of what we now call the Eastern trade) represented the extent of the English stake in the East. Bombay was the only post that England held in Asia. It was little considered, and the Crown sold it for a trifling sum to a trading company, soon after its acquisition, the date of that acquisition, as is perhaps unnecessary to state, being 1661, the same year as that in which England acquired her first post in the Mediterranean.

To acquire Tangier was in fact to come into direct relations with the East, for the Mediterranean was itself at that time far more a part of the Orient than it is at present. One may state this position even more strongly without over-stating it.

In 1661 the Mediterranean was oriental : to-day it is European. So that whereas the " Far East " was unexplored, and in great part undiscovered, the " East "—the influence of Asia and of Islam—was not only much nearer home than it is to-day, but was much more menacing. Not only was there a Turkish fleet, and a very powerful one ; but there were also the fleets of Tunis, Algiers and Tripoli, besides innumerable small piratical craft, which could not be traced or even watched, and which made commerce impossible to. conduct, except under the convoy of a fleet of men-of-war. It was the business of secret agents in the Mediterranean to keep the Admiralty informed, as well as they might contrive to do, of the existence, the movements and the strength of these fleets. It was the business of the governors of Minorca to keep on good terms with the Dey of Algiers, lest the corn supply should run short. The attitude of Europe was far more deferential towards Islam and her kingdoms than it is at present. Often it was supplicating; and sometimes vainly supplicating. Her intrusion on the East was haughtily tolerated. Her trade was admitted as an indulgence, and on terms that were ignored whenever it suited the East to remind the West that the sons of Islam were still lords in their own lands. This moderate statement applies only to the extreme limits of what, for the sake of a name, may be called the Khalifate. At the western end of the Mediterranean, at the point furthest removed from the sacred places of Islam, and from the sources of her spiritual and material strength, at the western end of the Mediterranean where England's first adventure was made, the situation was briefly as above stated. At the eastern end of the Mediterranean the situation was much more serious for Christendom. "The Empire," as it was then called (and which existed for another century and a half, growing yearly feebler), the Holy Roman Empire, stood face to face with the Turkish Empire, not as its descendant does to-day, but as a realm in mortal terror of a mighty and unscrupulous neighbour. On the Danube it was Islam that was powerful, united, menacing ; it was Christendom that periodically put on her armour, and composed her internal disputes in order to ward off, if might be, the blow that threatened her existence.

The dependence of the African fiefs on the Grand Seignior was merely nominal. But each of them was in itself too powerful for any European state to attack without extensive preparations, and the whole of Islam, with the exception (unimportant to Europe) of the schismatic kingdom of Persia, was knit into a semblance of homogeneity by the influence of her militant and victorious creed. Some estimate of the feelings with which England entered on her career in the Mediterranean may be formed from this short review. If on the one hand the forces of Islam were formidable and not patently disunited, on the other hand the countries which have since become Mediterranean Powers were at that time in a much inferior position, either in comparison with that which they now hold, or in comparison with the common enemy of all. Spain, under the influence of a great minister or a great king, an Alberoni or a Charles III., was still capable upon occasion of rising to the level of a first-class Power; and she habitually employed the language proper to a first-class Power, although it was becoming clear that her threats might often be disregarded. France was not yet a Mediterranean Power. Mazarin was only recently dead, the king had hardly asserted himself, and the star of Colbert had not yet risen. When the navy of France was built, it was rather towards Holland and England than towards the south that France looked for the enterprises that should occupy her seamen.

Italy (to use a phrase once current, but already thirty years out of date) was merely a geographical expression. Of the many small states that occupied the territory of the peninsula, only one, the kingdom of the Two Sicilies, was even of the rank of a second-rate Power. Nor was the navy of Naples very formidable; and when in the early years of the last century Murat sought the help of England, he was prepared to surrender the entire navy of his kingdom. Not even the genius of Acton could succeed in making the Neapolitans sailors. In the Adriatic Venice managed to hold her own against Turkey.

The first post that England occupied in Africa came to her from unfortunate hands—if she desired peace—for it was the last relic of the crusading empire of Portugal. For this reason a purely commercial policy was impossible in Africa. In Tangier, as in Bombay, commerce was desired, and English traders settled or endeavoured to do so. But in Tangier the native inhabitants were disinclined to accept the English as neighbours. Their persistent hostility could only be met by the maintenance of a large garrison, and Tangier developed into a military cantonment. This would have been impracticable in the case of Bombay, owing to the great distance of that settlement from England. In the case of Tangier it was not unwelcome to the Court to discover that a large garrison was indispensable for the commercial interests of the town. The difficult situation was dealt with by the Court in a manner sufficiently exceptional in the history of England to call for some notice.

In spite of occasional protests against the iniquitous occupation of Gibraltar, the time appears to be far distant when either the British Ambassador at Madrid will be directed to exchange the Rock for an equivalent, or when the Cabinet will be recommended by the Ambassador to offer to surrender Gibraltar as a means of facilitating negotiations with Spain. Both in England and abroad it seems to be understood that the Rock is as much a part of the British Empire as Portsmouth. Far other was the case for three-quarters of a century after Sir George Rooke captured Gibraltar. There was no clear conception of England as a Mediterranean Power, and Gibraltar itself was feebly garrisoned. Neither the sovereign, nor the Cabinet, nor the British Ambassador at Madrid had any continuity of policy in what now appears so important a question ; and from each of these three quarters there came, at different times, either offers, or actual undertakings, to surrender the Rock to Spain.

The first of these offers was made, not as a favour, not as the price of great concessions, but as a bribe to secure the fulfilment by Spain of treaty obligations. The offer was made by England on her knees, and it was haughtily and contemptuously rejected by Spain. Considering the relative maritime strength of Spain and England this is an astounding situation. It may be useful to review briefly the events which led up to the rejection by Spain of so favourable an offer. The war in the course of which Gibraltar was acquired, the War of the Spanish Succession, was brought to a close in the year 1713 by the Treaty of Utrecht. The bloodshed, misery and expense endured by Europe during the war were so serious, that the Great Powers were more than anxious to uphold the Treaty of Utrecht as a final settlement of the balance of power, especially of the balance of power in the Mediterranean.

There were signs, however, that one Power—Spain—was endeavouring to modify the existing situation for her own advantage ; and in the year 1718 four Powers—England, France, Austria, and Holland—entered into an alliance known as the Quadruple Alliance, for the purpose of maintaining the Treaty of Utrecht.

It seems an overwhelmingly strong combination, and by the Allies it was so estimated. There were only two Powers that were either likely or able to offer resistance to their will, and these were not very formidable Powers : they were the dukedom of Savoy, not very high up in the list of second-rate states, and Spain, decidedly low down in the list of first-rate states. Accordingly the Allies permitted themselves to use high language. The other Powers were to be given three months' time to join the Quadruple Alliance, and if at the expiration of that period they still declined to join it, the armed forces of the Allies were to be directed to compelling them to do so. In the meantime any attempt to modify the actual situation was to be considered a hostile act, and to be resisted by the united strength of the Allies.

They reckoned without the genius of Cardinal Alberoni,

who was at that time Prime Minister of Spain. When the Quadruple Alliance was signed, Spain had twenty-two ships of the line in the Mediterranean, and thirty thousand troops with the fleet; she was in actual occupation of Sardinia. Spain, who had long been decried in diplomatic and military circles as "effete," once more stood before the world erect, menacing, triumphant. Nor were her Mediterranean successes, great though they were, the principal part of her manifold enterprise. Alberoni aspired to nothing less than universal dominion. His diplomacy, as able as his war administration, had secured a firm alliance with Charles XII. of Sweden, a fighting monarch, who was not scrupulous as to the choice of his enemy so long as he was actively engaged in warfare. Alberoni determined that he should invade England. Russia engaged to supply the ships to convey the Swedish troops, and the Pretender was to land with these strange allies and enforce his claims to the crown.

In a panic of terror the British Ambassador at Madrid offered Gibraltar to Alberoni as an inducement to his master to join the Quadruple Alliance. Nor can he be blamed for making the offer. Spain was already mistress of the Mediterranean. Ships and armies, supplies and money, seemed to spring into existence at a wave of Alberoni's wand; England was bewildered at the spectacle of such resources wielded so energetically. The danger in the north was very real. The Cabinet was fully acquainted with the designs of Charles XII., and had only too good reasons for knowing that they were perfectly feasible. The dynasty itself was very insecure; and it was only three years since the Pretender had, without material assistance, convulsed the kingdom. On this occasion he was to repeat the attempt, but supported by Russia and Sweden, at a time when no help could be expected from the Mediterranean fleet, which could with difficulty hold its own in what had now become Spanish waters. No price could have been too high that would have secured the neutrality of Spain; the price actually offered was Gibraltar. This was at that time the greatest inducement that could possibly have been held out; for although England thought but little of the place, the Spaniards prized it most dearly. Had Alberoni accepted the offer, it is almost certain that England could never have become a Mediterranean Power; for the Rock could never have been re-taken if the defences and the garrison received even ordinary attention.

The country was saved from the consequences of its panic by the state of the Cardinal's health. Nobody but Alberoni could have lifted, unaided, the dead weight of Spain; but even on Alberoni that herculean task had told severely. In addition to the labours of the home administration, he had the anxiety of conducting a busy diplomacy over the whole of Europe, and his nerves were in a state of dangerous tension. His sudden and rapid successes, and the evident terror with which all Europe regarded him, coming on him while in this over-wrought temper, completely intoxicated him. What was the Rock of Gibraltar to a man who held Europe in his grasp? In imagination he saw his nominee, his creature, the Pretender, already enthroned at St. James'. When that was done he would take the Rock of Gibraltar, or anything else that he pleased. England was already on her knees; he would soon make her as one of his master's provinces.

These are the resolutions of a mind that had lost its balance; nevertheless, they are the resolutions on which the Cardinal acted. He rejected the offer, and continued his career of conquest in the Mediterranean by attacking Sicily. This was exactly the kind of move that had been contemplated by the Quadruple Alliance, when it was agreed that any attempt to disturb existing relations should be resisted by the united forces of the Allies. The only one of the Allies who could possibly resent it, however, was England. A British fleet was on the spot; the Admiral was Byng, created for his action on this occasion Viscount Torrington. He attacked the Spaniards off Cape Passaro and scattered their fleet; those of their ships that survived the battle fled towards Syracuse. It was on this occasion that Captain Walton, in command of the pursuing squadron, penned the despatch to his chief that has been often cited as a model of brevity :—

" SIR,
    " I have taken and burnt, as per margin ; going for Syracuse, and am, sir,
                            "Your obedient servant,
                                    " J. WALTON."

This occurred on August 22, 1718. In November the Duke of Savoy deserted his ally, Spain, and joined the Quadruple Alliance. The command of the sea being lost to Spain, the Austrians could pour in the troops set free by their recent successes against the Turks ; and Alberoni could not now hope to hold Sardinia and Sicily for long. In the north still worse fortune befell him. On December 11, 1718, Charles XII. of Sweden was killed by a cannon-ball at the siege of Frederickshall. With despairing energy the indomitable Cardinal got together and equipped another fleet for the invasion of England. In March 1719 it was scattered by a storm. In April the French, under the leadership of the Duke of Berwick, invaded Spain. The accumulation of disasters destroyed Alberoni's influence. In December 1719 he was dismissed from all his posts, and Spain subsided into the torpor from which the genius of her great Minister had dragged her.

Alberoni disappeared from Spanish politics ; but the idea of regaining Gibraltar for Spain did not disappear with him. The negotiations prior to the outbreak of war in the year 1718 had shown that, in the modern phrase, England was "squeezable" about Gibraltar. The Opposition in Parliament said a great deal about the "barren rock" and its "useless charge" to the finances ; and Continental statesmen, misunderstanding then, as often afterwards, the meaning of the attitude of a constitutional Opposition, presumed that they only had to press hard enough, and the Rock would be abandoned.

The Spanish king was a Frenchman ; but, following the wise counsels of Louis XIV., he had made himself a Spaniard of the Spaniards, and he pressed hard for the cession of a place so dear to Spain. He had a good ally in General Stanhope, the British Minister at Madrid; but not content with that, he made use of his family connection with France to engage the Regent to support his policy ; and the Regent's support was for the moment fatal to him.

On March 28, 1720, Stanhope wrote to Sir Luke Schaub a letter summarizing the whole situation. It ran as follows :—

"We have made a motion in Parliament, relative to the restitution of Gibraltar, to pass a Bill for the purpose of leaving to the king the power of disposing of that fortress for the advantage of his subjects.

"You cannot imagine the ferment which the proposal produced. The public was roused with indignation on the simple suspicion that we should cede that fortress. One circumstance greatly contributed to excite the general indignation, namely, a report insinuated by the Opposition that the king had entered into a formal engagement to restore Gibraltar. . . . We were accordingly compelled to yield to the motion, and to adopt the wise resolution of withdrawing the motion ; because if it had been pressed it would have produced a contrary effect to what is designed, and would perhaps have ended in a Bill which might for ever have tied up the king's hands. Endeavour to explain to the Court of Madrid that if the King of Spain should ever wish at some future day to treat concerning the cession of Gibraltar, the only method of succeeding would be to drop the subject for the present. We are much concerned that France should have interfered on this occasion ; the extreme eagerness that she testified was of great detriment."

As Stanhope was really anxious for the restitution of Gibraltar, he made a journey to Paris in order to persuade the Regent to withdraw his disastrous support; and the Regent, who had no desire for a breach with England, did so.

But Spain insisted. The king made a personal matter of it and a point of conscience. He had announced the impending recovery of the Rock to his subjects ; not that England was bound by so unjustifiable a promise, but it raised expectations in the minds of the Spaniards, thus forming a public opinion that grew more stubborn and angry as the months went by and the British flag still waved over Gibraltar. Stanhope therefore made a serious effort to get rid of the place. He was now Minister in attendance, and the sovereign was in Hanover. On October 1, 1720, he wrote a long despatch to England reviewing the whole European situation, and dwelling most emphatically on the importance of the support of Spain, and on the great desirability of not allowing the Regent of France to become too intimately allied with the King of Spain. Both these ends would be secured by the cession of Gibraltar. He could not, for reasons which he gave at length, advise the king to surrender Gibraltar for nothing, but if the King of Spain could give either Florida or Hispaniola in exchange, neither of which places was intrinsically valuable or of importance to Spain, either military or commercial, the Ministry would undertake to face Parliament with that proposal, difficult and perhaps dangerous though it might be to do so.

This was the second offer to withdraw from Gibraltar, and it was flatly and impatiently declined.

What stood in the way of the surrender of Gibraltar was the stubborn resolve of this country that the surrender should not be permitted. Nevertheless, so useless was the place considered to be, and so genuinely desirous were the king and his Ministers to shake this millstone from off their necks, that Parliament would undoubtedly have been again approached on the subject if Spain had been amenable. But it is hard to say which is the more curious—the conciliatory and almost imploring temper of England in a negotiation of such importance, or the authoritative language of Spain. England only asked a decent pretext for going ; but Spain would not be satisfied unless she not only expelled the English, but trampled on them.

Long disused as the Spaniards were at this time to the forms of constitutional government, they were probably really unable to understand why a project favoured by the sovereign, supported by his Ministers, and strongly recommended by his

ambassadors, could not be carried through. It is not unreasonable to suppose that Spain thought she was being trifled with.
Whether that was the reason or not, early in the year 1721
her Ambassador in London informed the Government that the
Court of Madrid was in real difficulty with the Spaniards,
owing to the long delay in restoring the Rock of Gibraltar,
and that they would be much obliged if England would make
the offer in writing ; it would greatly strengthen their hands,
it was added, with some *naïveté*.

Incredible though it may seem, the Ministers advised the
king to accede to the Spanish request ; and on April 29,
1721, George I. wrote an autograph letter promising to restore
Gibraltar for an equivalent. This was the third offer to restore
Gibraltar, and this time the offer was made in writing, and
over the king's own signature.

The British Ambassador at the Court of Madrid presented
this extremely compromising letter to the king and queen,
who were at the palace of Aranjuez. Very far indeed from
being gratified, they said that such a letter was useless to
them. Unconditional withdrawal from the fortress was the
only undertaking on the part of the King of England that
they would be justified in accepting ; it was what they were
entitled to ask, and the least that would satisfy the Spanish
people.

The fourth offer to surrender Gibraltar was therefore made,
also in writing, on June 1, 1721. The letter was subsequently
published in the *Commons Journal.* Translated it runs thus, as
to its material parts :—

" I do no longer balance to assure your Majesty of my
readiness to satisfy you with regard to your demand touching the restitution of Gibraltar, promising to make use of the
first favourable opportunity to regulate this article with the
assent of my Parliament."

George I. has been much blamed for writing this letter ; but
there are surely excellent reasons for justifying, and even
applauding, his conduct. He was not an Englishman, and

could not even speak English. He did not assume the arduous and complicated duties of a constitutional sovereign until late in life, and one of the first duties of a constitutional sovereign—as he was incessantly reminded—was to listen to the advice of his Ministers. The blame, if blame there was, must attach to Carteret and Townshend, who advised him.

There is no evidence that the Ministers resented the language held by Spain; there is no evidence that they cared anything for Gibraltar, either for itself or for what it might protect; indeed they maintained that it protected nothing except Minorca, which was quite capable of protecting itself. But if they ever felt that they were dealing somewhat lightly with a weighty matter, they reassured themselves by remembering the clause, "with the assent of my Parliament." From this point of view the letter, after all, meant nothing more than that they would take the vote of the House of Commons on the matter; and although for themselves they would have been very glad to get rid of the Rock, yet if it should turn out to be more valuable than they imagined, there was always the country to fall back upon.

The Spanish view was very different. The expression, "with the assent of my Parliament," conveyed nothing to a king who had no Parliament to consult. The letter was, in Spanish eyes, a sacred undertaking to evacuate Gibraltar made on the word of a king; and delay was now nothing less than perfidy. Accordingly, Philip, both personally and through his diplomatic agents, pressed eagerly for the immediate surrender. The British Ambassador at Madrid, early in 1722, wrote home bemoaning his helplessness. "Gibraltar," he said, "barred the way to all useful negotiation." He offered the suggestion that England might even yet obtain something in exchange. But Philip had from the beginning definitely rejected the idea of an exchange, and it is possible that the Ambassador anticipated no result from his despatch. Considering the view of the case that the Spanish king must naturally have taken, he exhibited great patience, and even the somewhat violent language with which he closed his last interview with Stanhope was not inappropriate. "Immediate restitution, or war," he said. Stanhope pleaded the old excuse about Parliament, and objected that the king was in Hanover. "Then let the king your master return at once from Hanover, and call Parliament for the purpose."

This was the last thing that the old king was likely to do, or his Parliament to consent to, and General Stanhope could promise nothing.

The Spanish king, strong in the consciousness of a recently concluded alliance with the emperor, then determined to take Gibraltar by force. On February 11, 1727, trenches were opened; and after much violent language the siege of the Rock was commenced. The garrison held its own without the slightest difficulty. There seems to be some doubt as to whether the Rock is impregnable as against modern artillery; but as against the weapons of one hundred and seventy years ago—especially of such weapons as Spain employed—there was no doubt whatever of the impregnable strength of the Rock. England retained the command of the sea, and easily fed the garrison, which numbered six thousand men. The siege lasted four months, and was from the beginning—as any expert could have foretold—perfectly futile.

In drawing up the preliminaries of peace after this outbreak, the Cabinet displayed its habitual anxiety to remove the British garrison from the Rock. The following letter from Lord Townshend to Stephen Poyntz, under date June 14, 1728, defines the situation :—

"What you propose in relation to Gibraltar (that is, the unconditional surrender of the Rock) is certainly very reasonable, and is exactly conformable to the opinion which you know I have always entertained concerning that place. But you cannot but be sensible of the violent and almost superstitious zeal which has of late prevailed among all parties in this kingdom, against any scheme for the restitution of Gibraltar upon any conditions whatsoever; and I am afraid that the bare mention of a proposal which carried the most distant appearance of laying England under an obligation of ever parting with that place, would be sufficient to put the whole nation in a flame."

Here then is a complete epitome of the situation. The Ministry and the Ambassador anxious, and markedly anxious, to abandon Gibraltar ; the king, as was his duty, showing no decided opinion ; and the Commons stubbornly (or, as Townshend put it, " violently and superstitiously ") opposed to the idea of withdrawal.

It was rumoured that an understanding on the subject had been secretly entered into by the sovereign. It was this—as became apparent from Stanhope's letter to Sir Luke Schaub —that had caused such general apprehension as long ago as the year 1720, fifteen months before the compromising letter was actually written.

The Opposition now angrily demanded that the king's letter should be tabled ; but Walpole demurred on the ground that the sovereign's correspondence was sacred.

The question was repeated early in 1729 in the House of Lords ; and as the royal letter had, in the meantime, been published on the Continent, it was difficult to avoid complying with the demand of the Opposition. The letter was tabled ; and now, eight years after the fourth offer to surrender Gibraltar had been made, Parliament was at last in possession of the whole case.

It was dealt with promptly and plainly. The least violent of several proposed resolutions was couched in the following terms : " That the House relies upon His Majesty for preserving his undoubted right to Gibraltar and Minorca." This was communicated to the Commons, and was ultimately carried ; although not until an attempt had been made to carry a resolution calling upon the King of Spain to definitely renounce his claim to both places. The moderation of the Ministry merits recognition. If any such resolution had been carried, a most embarrassing situation might have resulted.

Stanhope was now directed to present himself at the

Spanish Court and negotiate a peace. His task was comparatively easy. King Philip had endeavoured to obtain possession of Gibraltar by force, and had failed. The resolutions passed by Lords and Commons showed him plainly enough that he had nothing to expect from the English Parliament ; and although keenly resentful of the treatment he had received, he assented to the Treaty of Seville signed on November 9, 1730. In this treaty there were no provisions relative to Gibraltar. Considering the high language that had been held by the King of Spain for eleven years, ever since the fall of Alberoni in 1719, the omission was held to be equivalent to the tacit abandonment of Spanish claims to the Rock. The fifth offer to surrender Gibraltar was made in a secret despatch dated August 23, 1757. On this occasion William Pitt directed Sir Benjamin Keene, British Ambassador at the Court of Madrid, to offer Gibraltar to Spain as the price of a Spanish alliance to be contracted with the object of wresting Minorca from France. It is to be remembered that France had, in the preceding years, captured Minorca after a siege of seventy days' duration. Sir Benjamin's comment on this despatch was that Pitt must be mad, but he carried out his instructions, and made the offer, which was promptly rejected.

In contrast with the offers made thirty years before, it was on this occasion the Ministry that proposed the surrender and the Ambassador that derided it ; whereas in Stanhope's time the Ambassador had urged the surrender, and the Cabinet had shown reluctance, fearing the Commons. In spite of a sense of the grandeur and importance of the British Empire (a sense not only actual but prophetic), Pitt now sought to exchange Gibraltar for Minorca. So far was he from fearing that the loss of the Rock would weaken England, that he was fully prepared to employ his great authority in overcoming the resistance of the Commons to his proposal—a resistance that he must have known would be very stubborn. He, the Great Commoner, was perfectly ready to secretly dispose of what the Commons had repeatedly declared to be the most important possession of England, and to defend his action in the House afterwards. Nor is the attitude of Spain less noteworthy. Even after fifty years of alternate menace and negotiation Spain would not admit that Gibraltar was a proper subject for an exchange. Unconditional surrender was now, as before and after, the only offer that Spain would consent to receive.

The sixth and last offer to surrender Gibraltar was made in the year 1783, during the negotiations that preceded the Treaty of Versailles. Lord Shelburne offered Gibraltar to the Spanish Ambassador in exchange for Porto Rico in the West Indies. The Spaniard welcomed the prospect of regaining Gibraltar, but did not wish to cede Porto Rico in exchange, nor did he wish to cede any other place. When Shelburne communicated his proposal to his colleagues they were not scandalized ; they only said that his offer was too generous, and that England should demand two West Indian islands instead of one, in exchange for a place of such importance as Gibraltar. But if the Spaniards objected to giving up Porto Rico, still more did they object to giving up Porto Rico and Trinidad.

The sixth offer to surrender Gibraltar was made at the time when England was preparing to conclude a humiliating treaty. But on the other hand it was not contended, even at the time, that England had put out her full strength in the War of American Independence. Moreover, it was perfectly well understood that the American campaigns had been from first to last miracles of blundering, and the defeat of England was rather due to that cause than to any overwhelming military superiority on the part of the colonists. The nation was sullen rather than cowed. Moreover, England could claim glories. The English had held their own against the world on the sea ; and in particular the defence of Gibraltar had been one of the most magnificent defences of modern times, and had lasted for three years.

There have of course been Ministers who were insensible

to such considerations, and perhaps Shelburne was one of them. But even if he were he could never have made such an offer, if the maintenance of British power in the Mediterranean had been any part of a traditional English policy. However little the defence of Gibraltar may have appealed to the Cabinet, it stirred the nation to its depths. Shelburne tested the views of the Commons on the subject of the surrender, and obtained such unmistakable evidence that his proposal was unpopular that he was constrained to change his attitude. He informed the Spanish Ambassador that he would be unable to carry out his plan. The Court of Madrid was most indignant and threatened to renew the war. Thus once more in the course of the eighteenth century, Spain—eager as she was to regain Gibraltar—was yet unwilling to make the slightest concession to obtain her end, and showed herself most indignant that England would not give up the Rock for nothing and consider herself rather honoured than otherwise by the transaction. Shelburne's was the last offer made to surrender the Rock. These repeated offers on England's part to retire are, perhaps, worthy of attention for three reasons : firstly, because Gibraltar looms very large before the eyes of the student of the Mediterranean ; secondly, because—in so far as a civilian can understand the question—there can be no Mediterranean route to the East without Gibraltar as a starting-point ; and thirdly, because in spite of that fact it is not generally known how very carelessly England held this important possession for more than three-quarters of a century after it was acquired.

The conclusions to be drawn from this piece of history are these. The sovereign, as was his duty, held no very strong views on the subject. The Ministry and Ambassadors were almost invariably anxious to disembarrass British policy by ceding the Rock. Once only in the course of three-quarters of a century of negotiation was the Prime Minister deemed crazy for suggesting the withdrawal of England. Even the great Pitt saw no advantage in maintaining a British garrison in Gibraltar. He and the Ministers who were in favour of some other place in the Mediterranean had no idea of the Mediterranean route in their minds. They merely considered that Gibraltar might be useful to England as an offensive out-post. If any other place could be utilized for that purpose, any other place would be preferable. What baffled the plans of Ministers and Cabinets for shaking England free of Gibraltar was the stubborn resistance of the people, or, as Townshend expressed it, their "violent and superstitious zeal."

Gibraltar was not the first place that England occupied in the Mediterranean. Another place had been previously held and abandoned—one of far greater intrinsic value, and of incalculable commercial possibilities—Tangier.

Tangier is the gate to Morocco—a wealthy country woe-fully misgoverned. When the English retired they did so without entering into any diplomatic undertakings whatever; but inasmuch as it is now two hundred and sixteen years since they retired, their claim may perhaps be said to have lapsed with time.

To examine this page of history, why England went to Tangier, and what she was designing, is to unravel a very tangled thread, the clues of which have long remained obscure.

Tangier did not come to the Crown by conquest; it became English by the peaceful process of being included in a princess's dower—the princess being the bride of Charles II., Catherine of Braganza. There was another town that was acquired in the same way, and at the same time—Bombay: a place but little considered at the time in com-parison with Tangier. It was disposed of three years after its acquisition for a rental of £10 a year.

Both towns were fragments of the Portuguese Empire, once mighty, now fallen into decay. Portugal, though still main-taining her hold on Brazil, had renounced for the moment her ambition of becoming a world-Power. It was not un-natural therefore that she should surrender what she could no longer use to a Power that was now eagerly entering on a career from which Portugal was just retiring.

It is in respect of Tangier that the Mediterranean route is

first to be heard of. Not indeed from the sovereign or the diplomatic body or the Ministry—their views were all different, as will presently appear—but from the views of the merchants of the City of London. In particular the merchants engaged in the Smyrna trade were more than gratified at the guarantee of the safety of their vessels. The less important Indian interest was conciliated by the acquisition of Bombay. These were the reasons publicly put forward for the occupation of Tangier. They were sound reasons, and nobody inquired if others existed and were equally urgent. But the real reasons were concealed, and were much weightier than any that were publicly professed.

In reading pamphlets and private correspondence of the seventeenth and early eighteenth centuries one is inevitably struck with the almost unreasonable fear of standing armies displayed by the writers. At the present day the chief feelings on the subject may be said to be regrets that the army is not stronger and larger, but in the seventeenth century men wrote and spoke very differently. " To leave a prince his army is to bind Samson and leave him his locks." These and similar outbursts, decorated with all the pomp of notes of exclamation and capital letters, are the frequent expression of ideas that seem to have been entirely without basis.

But there were good reasons for much anxiety. Charles II. is generally accounted a mere trifler. That is what he became. But when he ascended the throne he had wide ambitions—bad ambitions they may be called to-day; but still they showed a great capacity for planning and scheming, which is not more remarkable on account of their disparity with the usually accepted view of the king's character than on account of their essentially un-English nature.

Briefly he desired to make himself master of England and an absolute monarch by means of a large standing army; and for that standing army Tangier was to be one of the depôts.

It is needless to say that these plans were not laid before

either the Ministry or the Houses of Parliament. They were confided to a secret committee of nobles, presided over by the king himself, supported by the Duke of York; the Duke of York was President of the Tangier Commission.

Charles was half a Frenchman by birth, and the great objects of his admiration were his grandfather Henry of Navarre, and his cousin Louis XIV. Louis did not altogether return this feeling; and before entangling himself in the wheels of his cousin's alarming adventure he waited to see how much the King of England could accomplish by himself.

This was in effect inconsiderable. Charles did not possess the vital force to be at once a great voluptuary and a great statesman of the Machiavellian type; and when he came to make his choice he preferred pleasure to politics. Moreover, his design ceased to be a secret—not improbably in consequence of his own indiscretion; his habits were too boisterous for a secret to be safe in his keeping. From the moment that his design became public property there was an end of its chances of success.

It is not until the outline, at least, of this tortuous and secret policy has been mastered that the failure of England in Tangier becomes intelligible. A knowledge of the king's ambition explains a great deal that went on there. Why for example should Lord Bellasis, who was a popular and successful governor, have suddenly resigned his post in exchange for the trifling appointment of Captain of the Pensioners?

The pay of Governor of Tangier was £1000 a year. This was the same as that of the Minister to the Hague, which was at that time a first-class Embassy, and was equivalent to at least £5000 or £6000 of our money. Moreover, there were large perquisites and commissions; and Lord Bellasis was not a man in the first flight by any means. But the transaction becomes plain enough when it is to be remembered that Lord Bellasis was on Charles's secret committee of nobles. In the event of the king's plan succeeding, he would become one of the first men of the kingdom, and in the meantime it was important for the king to have a man in England with local knowledge.

Again, why should a place which had been announced as one likely to be useful to trade, and which could in effect have been made most useful—why should such a place have been managed with such gross neglect in spite of the fact that the Duke of York—a good man of business—was President of the Committee? Because the duke only valued the place in so far as it furthered his brother's plans, and as those plans had now become impossible of execution, he took no interest in the town. Thus Tangier, which ought to have been well kept in hand as a military cantonment, was furnished with a mayor and corporation, a town council, grand jury and petty jury, a gallows, stocks, and a ducking-stool, as if to emphasize the fact that it was not to be a military depôt, but only an ordinary English country town. Again, the army, which was the only salvation in such a perilous position, was not only tied up with red tape, but positively strangled with it.

All this time millions were squandered in schemes never intended to be realized. The best governors were the dishonest ones, for they at any rate were well known at Whitehall, and could prevent stagnation ; the others could do nothing, except Kirke, who achieved something by disobeying orders.

So after an occupation lasting twenty-two years the English retired, having failed ignominiously and disastrously. The successive governors of Tangier received less than justice at the hands of Government. Several of them showed, in spite of everything, qualities that would have enabled them to make the best of the situation, and when the intrigues at Whitehall are considered, it is hardly surprising that so little was effected at Tangier.

Twenty years passed during which no post was occupied in the Mediterranean. In 1704 Sir George Rooke captured Gibraltar. In both cases it was apparently the people who clung to their acquisition, while the Ministry was indifferent.

As regards Gibraltar the sovereign was honest, as regards Tangier hardly so, and that is in part the reason why Tangier was perforce evacuated and Gibraltar retained.

These are the only two posts that England has held at the gates—one at the north, the other at the south of the entrance. Although the interest in Morocco has once more become actual in recent years, the British occupation of Tangier ceased to be a contentious feature in foreign politics more than two centuries ago. This is one reason why it calls for attention. It is to be considered that the history of the Mediterranean route to the East coincides with the modern history of England. In the days of the British occupation of Tangier the limit of the " East " was Smyrna. One hundred and forty years later Mr. Pitt could not bring himself to think of the Mediterranean except as an inland lake whose extremest point eastwards was Smyrna. Because the Smyrna trade had fallen off since the days of Charles II., Mr. Pitt considered that the Mediterranean was of no consequence. He said this immediately after Napoleon had endeavoured to reach India through Egypt, and had nearly succeeded. The influence of two distinct currents of thought is traceable throughout the whole course of this subject. On the one hand Ministers habitually regarded the Mediterranean as an inland lake where some trifling commercial interests were to be protected. They were therefore embarrassed by British conquests in the Mediterranean in proportion as these aroused the jealousy of Mediterranean Powers.

On the other hand, in spite of failing trade, the English people—not, one would say, an imaginative race—nevertheless found some reason for pushing along the Mediterranean. This was not a passing impulse, for it lasted for two hundred and thirty-six years. It was not damped by misfortune or even disaster. Nevertheless, the reason usually assigned for the presence of England occurred to nobody. There was some talk of the coral trade and some of sponges and fishery, while now and again there are allusions to the undeveloped timber resources of Sardinia. But it is hardly too much to

affirm that the advantages of our activity in this direction
were not demonstrable. The Mediterranean route to the
East is the one piece of foreign policy consistently pursued
by England for the last two centuries and a half.

The evacuation of Tangier in the year 1684 was entrusted
to Lord Dartmouth. In his last despatch from that place
he indulges in a prophecy. While lamenting the heavy blow
that England was dealing to her own interests in surrendering
the port to the Moors, he warned the Secretary of State that
the future held alarming possibilities for this country. Some
great Power must dominate the Mediterranean, and if it was
not to be England it would be France. He went further; he
ventured to insist that from his own observation France was
already working towards that end.

It was a reasonable ambition for France to cherish, and a
perfectly attainable ambition. At the time when Lord Dartmouth wrote, France was the only first-rate Power in the
Mediterranean. There was no such Power as Italy; Spain
was in complete decadence. Turkey was not formidable at
sea; Portugal had hardly recovered from the sixty years'
captivity, and had, moreover, resigned herself to a secondary
place among the Powers of Europe. There remained the
States of Barbary; and there could be but little doubt that
France would be a match for all of them combined, supposing
that they possessed sufficient political instinct to combine,
which hardly seemed probable.

The policy of converting the Mediterranean into a French
lake is a policy often promulgated, not only as desirable from
the French point of view, but as a sort of enforcement of the
rights of France. Until the last generation, however, there
have been but few successful steps taken in this direction, in
spite of the fact that it was a plainly declared ambition of
France as long ago as 1684. How is this to be explained—
that two nations are credited with pursuing a line of policy,
where the success of either must have excluded the other;
that one of those nations—always a first-rate Power, and at
the commencement of modern history by far the strongest

first-rate Power—advantageously placed for attaining her ends and with a definite policy clearly enunciated—this Power, France, made no progress worth mentioning for two hundred years; that the other nation, badly placed, badly directed, at first hardly a Power of the second rank—a mere pensioned dependent of her rival, should have pursued during all that time a course that has the appearance of a consistent policy, until she now dominates, in a measure, the waterway, from the gates of which she is separated by a thousand miles of ocean?

The explanation would appear to be this, that the French policy was an official resolution, a ministerial scheme, but it lacked driving power. The people were not behind it; it never was a national enterprise; a change of Ministry could always check, and the hostility of the monarch crush it. In England, on the other hand, while the Ministers were lamenting the embarrassments caused by the Rock of Gibraltar, and were employing every device to rid England of that incubus, the nation, incoherently perhaps, after the fashion of the English nation, but still stubbornly and resolutely, declined to have anything to say to the surrender.

The result was that England drove slowly and blunderingly, but still steadily, along the Mediterranean; while the French policy, with so many points in its favour, was yet pursued so languidly and fitfully that it can hardly be said to have been pursued at all.

Nothing better illustrates the contrast between these two tempers than the history of the British occupation of the now almost forgotten island of Minorca, the third British post along the waterway; and nothing better illustrates the extraordinary tangle of interests—national, religious, personal, and dynastic—out of which the Mediterranean route has grown, than the eccentric incidents which brought about the first capture of Minorca.

This took place during the War of the Spanish Succession; it followed and was caused by a serious disaster for England—the battle of Almanza. On this occasion the armies of

France and England, engaged in the Peninsula in supporting the causes of the French and English candidates for the throne of Spain, were commanded by two members of the House of Lords. The English army was commanded by a Frenchman, and the French army was commanded by an Englishman. The Frenchman was Ruvigny, Earl of Galway, a good soldier, well read in his profession, and of large experience. But his usefulness as commander of an English army was diminished by the fact that he spoke English very imperfectly. The Englishman was Fitzjames, Duke of Berwick, who seems to have possessed the tenacious courage of his father, the Duke of York, before the latter lost his nerve, and at least a spark of the genius of his uncle, the Duke of Marlborough. The English army was not only defeated but destroyed, and such of the survivors who escaped were very sore at their overthrow. As a means of occupying them, General Stanhope decided to employ them in the capture of Minorca.

If the French policy of dominating the Mediterranean had been seriously pursued during the quarter of a century that had elapsed since Lord Dartmouth evacuated Tangier, Minorca would have been found to be impregnable. It is quite reasonable to excuse the French for some lack of energy at first, even supposing their policy to have been a national one. For when England retired from Tangier, it seemed as if the only rival of France was removed, and as if she might pursue her Mediterranean policy unmolested.

No statesman could have foreseen the capture of the Rock by England. But when that event took place (and a most menacing event it must have appeared in the eyes of any statesman who desired a spirited Mediterranean policy for France) it would have been natural if France had fortified Minorca.

When Stanhope attacked it England and France had been at war for six years, and for four years England had held a threatening position at Gibraltar. A very little attention bestowed on the defences of Minorca would have made them impregnable. The fortress was naturally strong, and the harbour the best in the Mediterranean. The weak point of the island—the lack of supplies—was no great difficulty to a Power that was so near as Marseilles.

What actually happened was that the place was carried by a *coup de main* with slight loss. The French commandant was broken, and deprived of his Cross of St. Louis, which marked the anger of the king in unmistakable terms. But the net result of half a century of effort was this—that in a policy declared by French statesmen to be righteous, profitable, and essentially French, they had made no progress whatever. On the other hand, in so far as the policy of progress along the Mediterranean was definitely enunciated in England, it was derided. It was denounced as dangerous, unprofitable, un-English, in spite of which the progress eastwards continued. One strong position at the gates—Tangier —had been abandoned, but another, Gibraltar, far stronger than the first, had been occupied in its stead. In the year 1708 England had acquired a second fortress of the first class, which was within a hundred miles of the coasts of France and Spain. No wonder that the Courts of Versailles and Madrid grew uneasy. Nevertheless, at the Treaty of Utrecht, signed in 1713, five years after Stanhope's expedition, the island of Minorca was allowed to remain a British possession.

The capture of Sardinia is commemorated on Stanhope's tomb in Westminster Abbey, as well as that of Minorca. It is from Port Mahon, the principal harbour and fortress of the latter island, that Stanhope took the second title for his earldom. Sardinia looks the more considerable acquisition on the map. But no effort was made to retain it. Although it gave a royal title to the House of Savoy, it played but a small part in the history of the Mediterranean; and its politics did not again concern England until ninety years later. At the latter date, when England was in temporary occupation of Corsica, the project was put forward of occupying Sardinia in order to check the progress of the French Revolution. But it came to nothing, and the former occupation by England was hardly even remembered. In the correspondence of that time it can be readily discovered what was the temper of the country on the subject of Minorca. The keynote is struck in the letters of the Duke of Marlborough from his camp in Flanders.

MARLBOROUGH TO THE DUC DE MOLES.

"*Camp Terbanck,*
"*June* 26, 1708.

"I am working hard to get our squadron to winter in the Mediterranean. It would be of great public advantage for them to do so, but I find that our naval officers are of opinion that our battleships will be neither safe nor comfortable in Spezzia."

To the King of Spain to the same effect. } Same day.
To General Stanhope still more pressingly.}

MARLBOROUGH TO STANHOPE.

"*Camp Werwick,*
"*July* 15, 1708.

P.S. Autograph.—"I am so entirely convinced that nothing can be done effectually without the fleet, that I conjure you, if possible, to take Port Mahon."

MARLBOROUGH TO COUNT WRATISLAW.

"*Camp Werwick,*
"*July* 15, 1708.

"Everybody is agreed as to the necessity of having a winter squadron in the Mediterranean; but when all is said, we must defer to the judgment of the Admirals and naval officers on the question of the security of the port and other naval conveniences. Without doubt they are the best judges of such matters ; but I must confess to you that from all I can hear, these gentlemen think that the only possible place is Port Mahon. I have written to General Stanhope to do his utmost to take it."

To Count Sinzendorf.   Same date.

" If we only had Port Mahon all our difficulties would vanish ; I have already written to General Stanhope to do his utmost to take the place."

MARLBOROUGH TO THE MARQUIS DE PRIE.

" *Camp Fretain,*
" *September* 8, 1708.

" I have long been persuaded of the usefulness of our having a squadron in the Mediterranean during the winter. The only difficulty comes from our Admirals . . . but I have spoken so strongly on the subject that I flatter myself we shall yet succeed."

MARLBOROUGH TO THE KING OF SPAIN.

" *Camp Rousselaer,*
" *October* 24, 1708.

" We await impatiently the confirmation of the news that General Stanhope has succeeded in his expedition against the island of Minorca. We shall then be able to keep a good squadron of battleships in the Mediterranean during the winter."

To the same effect to the Duc de Moles.

MARLBOROUGH TO M. DE QUIROS.

" *Camp Rousselaer,*
" *October* 30, 1708.

" I heard from General Stanhope on the 30th of last month to the effect that the fortress of Mahon surrendered to him the day before." After compliments the letter continues, " There is no doubt that we shall now be able to keep a good squadron in the Mediterranean throughout the winter."

There is here no word of the Mediterranean route, no hint that Minorca is of any value, except in so far as in English hands it may prove to be obnoxious. There is no pretence, as there was at Tangier, that Minorca may prove to be useful from the commercial point of view, either for the commerce of the island itself, or for the protection it may afford to the commerce of other nations. It is an offensive outpost— offensive in every sense, as it enables England to carry the war nearer to France than would otherwise be feasible, and at the same time was intensely provocative to the French. The work of the Boyne was carried on in Minorca, which was regarded as a bulwark of liberty and a guarantee of the succession. It is exactly symbolical of the state of English opinion at that time that the first Lieutenant-Governor of Minorca was Kane, a veteran of the Boyne river.

England held Minorca uninterruptedly for forty-eight years. It came to be axiomatic that England should hold Gibraltar and Minorca in the Mediterranean. But the point of view has now somewhat changed since the Treaty of Utrecht, at which date Minorca was only looked on as a useful post to menace France. It has been noted how the people insisted on the retention of Gibraltar. It was the same in regard to Minorca. In both cases the popular interest was not more marked than official indifference. The state of men's minds was at first a state of intense gratification at the oppor- tunity of giving so much pain to France. This unworthy and futile state of mind was succeeded by another, not so clearly conceived or expressed, but pregnant with great possibilities.

In contemporary pamphlets and correspondence Minorca is constantly referred to as the glory of England. But if this nebulous phrase be condensed it resolves itself into the conclusion, that Minorca is necessary as a support for Gibraltar.

In the alternative statement of this position Gibraltar is described as the brightest gem in England's crown. But the orators who so describe it always end by concluding that Gibraltar is necessary as a support for the island of Minorca. Each place is confessedly useless in itself, and though there is no reason for holding either, it seems indispensable to national honour to hold both.

Officially both places were treated negligently. As regards Gibraltar the popular mind was so firmly made up that it

D

was unnecessary to make a demonstration on the subject. Moreover Gibraltar was admittedly impregnable, so there was no cause for anxiety, though much for discontent. But as regarded Minorca things were different. It was not by any means certain that Minorca was impregnable, and consequently there are frequent complaints of the insufficient strength of the garrison, and Parliamentary inquiries as to the negligent conduct of affairs there.

It is no part of this subject to inquire into the circumstances of the first siege and fall of Minorca, which took place in the year 1756, any more than to trace the commonplace and obscure history of the administration of the island during the half-century from 1708 to 1756. The important matter is the temper of the nation on the subject, and the growth of an idea of a route to the East through the Mediterranean. There have been few more remarkable outbursts of national anger than that which followed the fall of Minorca. The feeling was quite unmixed with the idea of extenuating the guilt of Byng. There was a howl of wrath, and a demand for revenge and nothing more. This is a very grave state of mind for England to find herself in. It is the exact equivalent of the state of mind in which the French exclaim, " Nous sommes trahis." On this occasion they might have said so with some reason.

Rightly or wrongly the nation had for half a century past insisted that England was now a Mediterranean Power, and that it was the duty of the Ministry to see that her authority and influence in the Mediterranean remained unimpaired. Over and over again had they been promised reform. Select Committees of the House had been appointed and evidence taken, but nothing done. The people felt that their Ministers had deceived them ; and the extraordinary slowness and inefficiency of the preparations for the relief of the fortress, culminating as they did in an ignominious repulse, produced a state of mind only to be soothed by bloodshed. Byng, the Admiral in command of the relieving squadron, was marked out as the victim ; an innocent man, but that was of no consequence. Byng at any rate was within their reach, if the Ministers were beyond them. It made no difference that Pitt, the people's favourite, stoutly declared that Byng may have made a mistake, but did not deserve death. If Pitt had offered them another victim they would have released Byng, but he did not ; and the idea that Minorca was gone and nobody was to hang for the loss was intolerable. So Byng died.

It was immediately after this incident that Pitt made the fifth offer to surrender Gibraltar to Spain, in exchange for her help in recovering Minorca for England.

It is difficult to catch the drift of Pitt's policy with regard to the Mediterranean. Probably he had none, and only sent his orders to Sir Benjamin Keene with the object of winning the favour of the people by recovering Minorca. But that he should have been willing for that purpose to trade away Gibraltar, a place which the people were stubbornly resolved to hold, shows how little attention he had given to the subject. The fact would appear to be that Pitt was ambitious of making England a great colonial and Indian empire. He was not particularly ambitious of making her a Mediterranean Power, either for the reason that influence in the Mediterranean was in itself desirable, or for the reason that the Mediterranean was a valuable route to the East.

Nevertheless immediately after Sir Benjamin Keene had called him a madman for offering Gibraltar to Spain (and with some reason, it must be admitted), Pitt did make an attempt to secure for England a second post in the Mediterranean, and his negotiations are thus described by Lord Chesterfield :—

> " *Blackheath,*
> " *September* 30, 1757.
>
> " I am told, and I believe it is true, that we are negotiating with the Corsican—I will not say rebels, but assertors of their natural rights—to receive them and whatever form of government they think fit to establish, under our protection, upon condition of their delivering up to us Port Ajaccio, which

may be made so strong and so good a one as to be a full equivalent for the loss of Port Mahon. This is, in my mind, a very good scheme; for though the Corsicans are a parcel of cruel and perfidious rascals, they will in this case be tied down to us by their own interest and their own danger—a solid security with knaves, though none with fools."

At this date—the middle of the eighteenth century—France appears to have at last awakened to the fact that her Mediterranean policy must be more actively pursued unless England is to be paramount. Not that the English were alarmingly encroaching; and in particular the plan mentioned by Lord Chesterfield came to nothing. But it is noticeable because the time was coming when England was to have a good deal to do with Corsica. This island was an ancient possession of the Republic of Genoa. At the time when Lord Chesterfield wrote, she was struggling to throw off the yoke of Genoa, and declare herself independent, under the leadership of Pasquale de Paoli. Postponing for the moment the study of Paoli and his struggles with the Genoese, it may be well to consider the drift of the Seven Years' War.

This war lasted from 1756 to 1763, and was concluded by the Treaty of Paris. As the net result of the war, France suffered heavy losses on the West Coast of Africa, still heavier losses in India, and was expelled altogether from Canada. She regained some of her lost ground in the West Indies, but on the whole her colonial empire suffered most severely.

What bearing have these events on the Mediterranean route? This, that the expansion of France, to use the modern phrase, having received such a serious check abroad, the question arose whether she could not expand in Europe. She had been compelled to cede Minorca to England at the Treaty of Paris, and in her search for a counterpoise, she fixed on the derelict island of Corsica, the acquisition of which served a double purpose. It partly compensated France for her colonial losses, and it strengthened her near home. It is to be observed that even yet—in 1768—there is no notion of the Mediterranean as a route that leads any-where. France was alarmed at the presence of England, not because she recognized that the Mediterranean was a link in the British Empire that she was desirous of breaking, and profiting by the fracture, but because she imagined England to be animated by the sole ambition to injure her. Corsica was *de jure* Genoese ground ; *de facto* it was the private estate of Pasquale de Paoli. In the year 1768, five years after the signature of the Treaty of Paris, France purchased the island of Corsica from the Genoese Republic, and sent in armed forces of such strength that there was no hope of resisting them. The island became a French province, and Pasquale de Paoli came and settled in London.

In the south aisle of Westminster Abbey there is a bust by Flaxman of Pasquale de Paoli. His peevish features represent him as he was known to Londoners in the early years of the last century. The eulogium which is en-graved beneath the bust refers flatteringly to his achieve-ments in early life, or what is somewhat woundingly called "the earlier and better" part of his life. It appears to be insinuated that the latter part of his life was not entirely creditable to him. This painful silence refers to his behaviour during the English occupation of Corsica. At the close of the Seven Years' War England resumed the position she occupied by the signature of the Treaty of Utrecht, and held Gibraltar and Minorca. These two places were retained for nineteen years. Five years after the restoration of Minorca to England, France made a return thrust by occupying Corsica. The effect of this measure was to quicken the nation's appreciation of the places held by England. It even raised some slight uneasiness. "Corsica a French province is terrible to me," said Burke. But it was not terrible to any one else in the House except Admiral Sir Charles Saunders, an authority no doubt. It may be added that the French hold on Corsica has always been singularly inoffensive ; and the instinct of the English nation which impelled them to do nothing more than quietly watch the French there, was perfectly sound. England therefore rested content with Gibraltar and Minorca, and remained there unthreatened in spite of the French occupation of Corsica. The story of the second period of administration of Minorca by England has no particular interest, but as the time of occupation draws to its close, the student is confronted with the first combined and determined effort on the part of France and Spain to expel England from the Mediterranean altogether. Both fortresses were attacked at once. Minorca stood a siege of one hundred and seventy days, following on a long but loose and inefficient blockade, and then surrendered.

It was expected at the time that the fall of Gibraltar would immediately follow ; for it was commonly said that the two places were interdependent. In time of peace or of a loose blockade they certainly were, but only to this moderate extent—that the comfort and convenience of each place was largely subserved by the possession of the other. But when all thoughts of comfort and convenience had been long since laid aside; when the enemy had for the moment the command of the sea, and both places were strictly invested, there was no reason why the fall of one place should entail the fall of another. The French reckoned a great deal on the British being discouraged, and on a loss of *morale* in the garrison of the place that still held out. Such reckonings proved futile. A real and substantial advantage gained by the fall of Minorca in the early part of the year 1782, was that the besieging army and fleet were set free for the operations before Gibraltar. These were on a grandiose scale.

The two Bourbon kingdoms of France and Spain signed a treaty of offensive alliance on April 12, 1779 ; and war was declared on June 16 ; both events took place at the height of the disastrous quarrel between England and the American Colonies, soon to be the United States of America.

No peace was to be made without the restoration of Gibraltar to Spain. As regards Minorca, it was only provided that every effort was to be made to recapture that place ; and the French and Spanish efforts in this direction were ultimately successful.

But in the objects of the alliance Gibraltar takes the first place and Minorca the second place only. This is in strong contrast with Pitt's policy, according to which it was worth while to buy Minorca at the price of Gibraltar.

The complete investment of Gibraltar soon brought the garrison to straits. A relieving force was fitted out and entrusted to Rodney, who sailed on December 29, 1779. He not only threw supplies into Gibraltar and Minorca early in January 1780, but succeeded in severely beating a Spanish squadron of eleven ships, of which he captured or destroyed seven in an action off Gibraltar fought on January 16, 1780. On December 20, 1780, war was declared against Holland, and England was now confronted with four great nations in arms, three of them—France, Spain and Holland—being naval Powers.

On April 12, 1781, Gibraltar was for the second time relieved. But on February 5, 1782, Minorca fell—her garrison being reduced to 560 men, or less than one-twentieth the force of the enemy. The island had held out for six months, and the defence was a most gallant and brilliant performance ; but its fame has been lost in the splendour of the great defence of Gibraltar.

By September 10, 1782, the famous Rock had already been besieged for three years, and although twice relieved its garrison was suffering considerable privations. The fleet hitherto engaged in the siege of Minorca having been set free by the fall of that place, was now able to join the fleet besieging Gibraltar, and the united squadrons numbered fifty sail. The garrison numbered seven thousand men ; the besieging armies no less than thirty-three thousand, or more than twice the largest number ever occupied on the reduction of Minorca. Besides the fire of the fleet, there was concentrated on the Rock the fire of three hundred pieces of artillery on the isthmus, and one hundred and fifty-four heavy guns on floating batteries ; there were also forty gun-boats and forty bomb-vessels. The grand attack was made early in September 1782. It was to have been succeeded by a grand assault, for which, however, the opportunity never occurred, as the grand attack failed. Six thousand five hundred cannon-balls and eleven hundred shells were discharged against the Rock during every twenty-four hours for four days in succession, and then the floating batteries were moved into action. The siege took an aspect of almost mythic grandeur ; and the prize at stake was the Mediterranean route to the East.

The details of the defence are of secondary importance. The efforts of the assailants were on so gigantic a scale that they provide a just measure of the heroism that defeated the attack. The attack failed. It was followed by a mighty fleet action, the object of which was to beat off the relieving squadron—the third relieving squadron during the siege— under Lord Howe. The fleet action ended in favour of the English. It was plainly impossible to starve the place out. It was also plainly impossible to take it by force. From that moment to the present Gibraltar has been, with the exception of Lord Shelburne's offer, an unchallenged British possession. This, the sixth offer, was made immediately after the great siege—an astounding example surely of the Olympian imperturbability of the diplomatic temper. For twelve years England remained passive in the Mediterranean. Gibraltar she still retained, but without much idea of securing a support for the Rock. Aggressive schemes were no longer possible, and she was contented to make the most of her shattered resources. In the year 1789 the French Revolution broke out, and the effect on the Mediterranean was immediate. Pasquale de Paoli quitted London, was elected to the Convention, quarrelled with the Revolutionists, fled to Corsica with a price on his head, roused the country to arms, and stood at bay.

In the meantime England had been dragged, much against her will, into the Revolutionary quarrel. She determined to uphold the Bourbon cause, and to that end occupied Toulon.

This was a great feat of arms. Thirty ships of war of seventy-four guns and upwards, or more than one-third of the line of battle force of France, were captured, with more than twenty frigates. But this first success was fruitless, for the country did not rise in favour of the Bourbon party, as had been anticipated.

The aim of England had been to form a nucleus of resistance—a point round which the Royalists might rally if they were so disposed. This was effected. But the Royalist movement was not powerful enough to hinder the investment of the city from the land side. A somewhat ridiculous investment it was at first, with Casteaux the painter, and Doppet the physician, for its generals—men who knew as much about the range of artillery as might be expected from distinguished artists and doctors.

Napoleon Bonaparte, however, supplemented their ignorance. In 1793, at the age of twenty-four, he achieved his first considerable success by driving the English out of Toulon. They were not even able to carry away all their prizes, and fifteen ships of the line remained behind in Toulon harbour. These afterwards formed the nucleus of the fleet that conveyed Bonaparte to Egypt three years and a half later.

The immediate result of the recapture of Toulon was the occupation of Corsica by the English.

This is sometimes ascribed to the sagacity of Admiral Hood, who saw the potential strength of the island, and sometimes to the diplomatic ability of Sir Gilbert Elliot, who negotiated the surrender. But in point of fact the mainspring of the whole movement was Pasquale de Paoli. The only quality that this eminent man possessed in a superlative degree was the power of impressing other men with his own importance. He was a good orator of the ornate style, a fair swordsman, a good organizer of a half-civilized state, a good leader of guerilla warfare. But these qualities will not of themselves account for his high reputation. While in London he had enjoyed what was at that time the ample income of £1200 a year as a pension from King George III. He addressed the

sovereign direct : his letters are very stately productions. One
anticipates the commencement, " Sire, my good Brother."

In 1793 there is no doubt that he was for the moment
master of Corsica ; but he could not hope to hold his own
for long against France.   He therefore in an autograph letter
invited King George III. to take possession of Corsica, and at
the same time he invited Hood to commence the good work
of conquest forthwith.   His social position in London made
his request for help almost authoritative; but the commis-
sioners referred to London before embarking on their enterprise.   They received in reply one of the most embarrassing
sets of instructions ever issued to a plenipotentiary.   The
instructions amounted to this, that neither the king nor the
Minister, Mr. Dundas, knew anything whatever about Corsica ;
or had any hopes of finding out anything that would be likely
to be useful to Sir Gilbert Elliot, the Civil Commissioner.
Elliot was directed to steer his own course.

The plain truth of the whole transaction is that England
conquered Corsica to oblige Pasquale de Paoli, and was subsequently turned out because Paoli was not made viceroy.
That is the true history of the three years' occupation.

But one incident deserves a slightly longer notice.   Corsica
was the only one of British Mediterranean strongholds (with
the exception of Tangier) where any attempt was made to rule
constitutionally.   In all other cases military law prevailed.   In
Corsica a constitution was introduced, modelled on that of the
United Kingdom and establishing trial by jury.

Trial by jury suits English people for various reasons—
racial, social and historical.   But although in moments of
fervour it is acclaimed as the palladium of our liberties, there
is no reason why it should be presumed to be efficacious as a
remedy for the ills of other nations.   That presumption—not
always a safe one—was made as regards Corsica.

The *vendetta*, by the laws of which every man was bound
to support his friends and to slay his and their enemies, reduced
the administration of justice to an absurdity.   It could hardly
be otherwise in a place where the foreman, or even the judge

himself, might be the hereditary foe, or perhaps the cousin, of the prisoner, and where a very large sympathy for crime existed in all classes of society. So the preamble to the first Act of the second Session of the Corsican Parliament ran thus :—

" Considering that the institution of trial by jury has hitherto favoured the immunity of crime," trial by jury is abolished.

This incident, curious in itself, had a marked effect on the Mediterranean route by weakening the hold of England on the hearts of the Corsicans. It more than doubled the difficulties of internal administration, and led directly to the conviction that the occupation was a mistake, which it was not from the strategic point of view. Strategically it was an act of great wisdom. Twice during the wars of the French Revolution the command of the sea threw into British hands the power of completely checking the progress of the French in Italy. The first opportunity was afforded by the occupation of Corsica ; the second by the occupation of Sicily. Of the second opportunity England made good use, and with the best results. From the military point of view there may be said to have been no mistakes, partly perhaps because she had profited by experience. But the first opening was thrown away in the most lamentable manner.

Practically in the year 1795 France had one enemy in arms on the Continent—Austria. Austria could be attacked on two sides by Germany and by Italy. In Germany nothing could be done. But the attack on the Austrian possessions in Italy must be made by the Cornici road ; and one need not be a military man, one need only to have travelled along the Riviera, to recognize that an attack along that line is perfectly impossible to any army that is not covered by a fleet. More, the artillery and stores for the use of the army of Italy were mostly conveyed by sea from France, and landed wherever possible, at Alassio and elsewhere along the coast. From the secure position in Corsica commanding the sea-way, England could give victory to either side.

It is painful to recollect that the Austrians, relying on British promises of support, did advance to somewhat perilous positions, and that they were heavily defeated. Guns and stores went unchallenged by sea, and the French fleet even supported the French armies in spite of the presence of a greatly superior British squadron. The passage being once forced and a French army in Italy, the opportunity was gone. In the next year Bonaparte took command of the army of Italy, the whole Peninsula was subjugated, and the English driven out of the Mediterranean.

And yet Nelson was there; but not commander-in-chief. At the age of thirty-eight he was still a captain commanding a sixty-four, the *Agamemnon*. He did all that he could; had his chief been Hood he would have done much more. Had his chief been Jervis he would have done everything. Jervis, who assumed the Mediterranean command at the end of 1795, ranks second to Nelson only amongst British admirals, and was to win his earldom in 1797 for the battle of Cape St. Vincent.

Hood had done well in the Toulon command. But he was seventy-five years old, and the reduction of Corsica had greatly tried his strength. In November 1794 he had gone to England, and the command devolved on Rear-Admiral Hotham. Between Hood and Jervis there is thus an inter-regnum of Hotham lasting for one year—from November 1794 to December 1795. It is hardly possible to over-estimate the damage caused to England by this one year of Hotham's command.

Byng, Parker and Hotham were all leisurely men. Byng did but little harm, but paid for his sluggish conduct with his life. Parker received only a gentle and implied censure for his share in the battle of Copenhagen. Hotham escaped censure altogether, although his behaviour was little short of monstrous.

In 1795 the French were still disposed to try fleet actions with the English. They had not yet abandoned that policy for the alternative policy of commerce-destroying. Martin was the Admiral in command at Toulon, and on March 2, 1795, he put out from Toulon with a fleet of twenty-seven sail. On the 11th and 12th the English and French were face to face. No doubt the English fleet was badly provided ; but of Martin's twelve thousand men nearly two-thirds were raw hands. It is not wonderful, therefore, that the English were victorious in the action that ensued. What is wonderful is that in the moment of victory the Admiral should have stayed his hand. Nelson did his best, and urged him to follow up his advantage, pointing out that with a slight exertion he could now destroy the French fleet, and irreparably damage them in the Mediterranean. But Hotham had no intention of damaging the French. "We must be contented," he said. "We have done very well." This was gall and wormwood to Nelson, who, like all great conquerors, thought nothing done while aught remained. The Admiral, he said, "had no head for enterprise, perfectly satisfied that each month passes without any losses on our side." This is a very proper temper for an admiral at anchor in the Downs, with nothing more important on hand than social engagements ashore, and the observance of daily routine. But in warfare it is dangerous ; with a Napoleon to deal with— disastrous. The home authorities can hardly be blamed for the failure. No single Cabinet Minister, still less a whole Cabinet, had ever formulated a Mediterranean policy for England. As for the people, even in time of peace they had been unable to do more than insist that England ought not to retire. In 1795, with the storm of the Revolution raging about them, they were far too bewildered to afford any light of counsel to the Cabinet. Everything was left to the man on the spot, and the man on the spot was Hotham.

When Jervis relieved him in December 1795 the damage was done. The French were already in Italy; the Italian States were thoroughly discouraged, and Corsica was heaving with discontent. England was now on the defensive. In July 1796 Elba was occupied; the next month an alliance was signed between France and Spain; England replied by seizing Capraja. These energetic measures, if taken a year earlier, might have been profitable. Taken in 1796 with forty hostile sail to deal with, and only fourteen sail to meet the danger, they were hardly more than a piece of feeble brag. Early in 1797, in the greatest agitation and alarm, England evacuated all three islands and concentrated on Gibraltar. The Mediterranean was freed from the English.

Thus, one hundred years ago that state of things had been reached in the Mediterranean which had long been desired in France, and also by some parties in England; the English had retired from the Mediterranean, or, as the French would have put it, and with good reason, they had been expelled from the Mediterranean. Whichever word be selected, the fact remains—during the year 1797 not an English ship entered the Mediterranean. Gibraltar indeed was still held, but although three definite attempts had been made to extend British influence with Gibraltar as a base, three heavy failures alone had resulted; and the last and greatest of them had not only been disastrous but ridiculous. Gibraltar in English hands might for the future be regarded as harmless; it did not materially alter the position of the Mediterranean as a sea in which French influence was paramount. What ensued? What was the immediate result of England retiring from the Mediterranean? The immediate result was that France made her famous dash on the East. This had already been prepared for more than a year past. When the Venetian Republic was broken up, France retained Corfu as a good recruiting ground for sailors, a first-class fortress, and a safe harbour for the French fleet to shelter in, if forced to quit the open sea by superior forces. "With Corfu and Malta we shall be masters of the Mediterranean," said Napoleon. Corfu was already in his hands; Malta he was to seize in the course of 1798. His preparations were hurried on as fast as possible, and at the close of the year 1797 were nearly completed. The expulsion of the English enabled him to use his preparations to the greatest possible advantage. His object was nothing less than the domination of the East.

An earlier generation was accustomed to hold in somewhat inflated language that England was the great object of Napoleon's hatred, as being the only Power capable of standing up to him. Then came a change; it was pointed out how gigantic were the land operations of Napoleon; how comparatively small the military efforts of England. There was a reaction in the Emperor's favour; it began to be held that England had been purely malicious towards France. "What," it was said, "had Napoleon to gain from opposing England?" What on the other hand had she to gain by opposing him? It was merely an exaggerated sense of the importance of England that had caused men to look on themselves as especially marked out for Napoleon's enmity. England would have done much better to have selected him for an ally. Later and fuller knowledge proved that the English of 1800 were right after all. The vast land operations that loomed so large in the eyes of the English of 1810 were all directed in reality against England. This appears in Napoleon's own writings; from such expressions as, "I will conquer the sea by the land;" "I will reconquer Pondicherry on the banks of the Vistula."

What he coveted was the East; he could not attain his object, because England stood in the way; therefore he must destroy England. The first means that he employed to that end was to attack her fleets, and when that failed he aimed at the destruction of her commerce. This was to be attained by the monstrous commercial system set forth in the Berlin Decrees and in the Milan Decrees. This system, thrust upon Europe at the point of the sword, produced that general revolt of the nations which ended in his vast European campaigns. But these campaigns, though apparently fought against Prussia and Russia, were in reality so many blows aimed at England.

It may be noted here that Napoleon's view was erroneous. He thought that England was rich and powerful because she held India. The contrary was the case. He mistook cause for effect.

But when all is said, the East was from the first his objective, his youthful dream, his darling project through life; and he set out on his Eastern campaign perhaps with some memories of Cortes and Pizarro in his head, but with more of Alexander.

The way being clear and all things ready, whom did the young Napoleon take with him?

Desaix	Lannes
Kléber	Andréossy
Murat	Caffarelli
Berthier	Belliard
Davoust	Marmont
Reynier	Junot

—the very pick and flower of the army of France. Of these Murat was afterwards Grand Duke of Berg and King of Naples, Berthier Prince of Neuchatel, Davoust Duke of Auerstadt and Prince of Eckmuhl, Lannes Duke of Montebello, Marmont Duke of Ragusa, Junot Duke of Abrantes, Desaix died at Marengo, Kléber was assassinated in Egypt. Alexander must have Alexander's generals with him; and the brilliant destiny of all these young men was really nothing to what Napoleon was imagining for them as they sailed to Egypt together. Instead of Grand Dukes of Berg, or Princes of Neuchatel, they were to be Satraps of Persia and Bactria, Sultans of Roum, Syria or Nubia, and the young Napoleon would have been their Caliph.

"I ought never to have come back from Egypt," he afterwards mournfully said; but in point of fact there was no choice for him.

As to his objects—already in the previous year he had written, "The time is not far distant when in order to destroy England we must make Egypt ours."

When once under way he stated plainly that the object of the expedition was to strike a death-blow to England. It is to be observed that his plan was a more important operation than a mere experimental raid. It comprised the colonization of Egypt—he mentions the number of families required in order to give France a firm hold on the country (forty thousand)—the cutting of the Suez Canal, the conquest of Syria as the indispensable basis for his operations against India, and the rescue of the relics of the French power in India by the simple process of overthrowing the English there.

Napoleon has been severely blamed for his expedition to Egypt, because, it is said, he showed such an ignorant underestimate of the naval power of Great Britain. It is true that he habitually under-estimated his great enemy in this respect, and still more under-rated the commanding influence conferred by the navy. But if he was ever justified in believing that he might neglect the sea-power of England, he surely was so justified in the year 1797.

In order to seize the East the first and indispensable measure was to drive the English out of the Mediterranean. That was already achieved. The next step was to hold their fleets occupied outside Mediterranean waters. That also was achieved. The powerful Spanish fleet was regarded not only by Napoleon, but by English observers, as more than a match for what was left of the so-called Mediterranean Squadron. In the north the Dutch fleet kept the Channel Squadron pinned down in the narrow seas. A French invasion of Ireland was attempted. Consols fell to 51 ; everywhere in the United Kingdom anxiety and distress prevailed. Keeping steadily before the eyes of the public the menace of his invasion to England, Napoleon pushed on the preparations for his Eastern campaign in perfect secrecy and with as much despatch as was possible in the state of the dockyards.

But the temper of England, steady and redoubtable in this hour of danger, had nearly snatched away the opportunity before even Napoleon could use it. On St. Valentine's Day, 1797, Sir John Jervis with fifteen sail attacked the Spanish fleet of twenty-seven sail off Cape St. Vincent. When the fight was over, the battered Spanish fleet drew off, leaving four of her largest vessels prizes in the hands of the English.

In October of the same year Duncan gained a yet more

E

glorious victory over the Dutch fleet off the Texel, capturing
nine vessels out of the enemy's fifteen. "Gentlemen," said
Duncan to his captains before the action, " I have taken the
soundings of the Channel, and I find that if the flagship is
sunk, my flag will still fly above the water." Such was the
temper of England in 1797. Yet so serious was the crisis
that even after the battles of St. Vincent and the Texel,
Collingwood could write that it was still a question whether
the English were to be any longer a people. Nevertheless
much had been done and the Mediterranean claimed
immediate attention.

On April 10, 1798, when Sir Horatio Nelson sailed
from England to join the Mediterranean Squadron, Bonaparte
had not yet quitted Paris. On May 2, Nelson sailed
into the Mediterranean to reconnoitre. The next day Bonaparte started for Toulon to take command of the army of
Egypt. On May 19 he sailed, and on the 20th a violent
gale drove him to take refuge in Genoa, and Nelson to take
refuge in the S. Pietro islands at the south of Sardinia.
Nelson reached this harbour dismasted on the 23rd. Three
days later Bonaparte passed him at sea, having sailed down
the Italian coast inside the islands of Corsica and Sardinia.
On June 9 the French were before Malta, and on the
12th the place capitulated. Thus was achieved without loss
the third part of Napoleon's great scheme. And yet with
victory apparently within his grasp, he was already defeated.
Notwithstanding the rapidity and the unparalleled secrecy of
his movements, he was after all only a week ahead of Nelson.
On June 12, 1798, Malta capitulated to the French.
Five days earlier, Trowbridge joined Nelson, and the
Mediterranean fleet numbered thirteen seventy-fours and one
smaller ship. From this date the movements of Nelson are
but the wheelings and circlings of a hawk before he swoops on
his prey. Nelson's instructions contained no hint (as how
should they, being drafted in ignorance ?) of the real destination of the Toulon fleet. They told him that a descent on
Ireland was contemplated, or a descent on the coast of Sicily,

or a descent on the coasts of Spain or Portugal ; there was no mention of Egypt.

Bonaparte then sailed south for Malta. Nelson sailed in the opposite direction, made his way round the north of Corsica, through waters already only too familiar to him, and made Civita Vecchia on June 14, two days after the surrender of Malta to the French. In consequence of information received at Civita Vecchia, he struck out of the possibilities of his instructions both Ireland and Spain and sailed south. On the 17th he made Naples, and on the 20th Messina. Here he heard of the capture of Malta.

Nelson always supposed that Bonaparte coveted Malta, because it was a good point of attack for Sicily. He was astounded to learn that after taking Malta the French had left Sicily alone and sailed east. But his first movement of astonishment was succeeded by an inspiration of genius— Egypt. It flashed across his mind, Egypt and the East must be Napoleon's destination. He crowded all sail and made Alexandria on June 28, and found nothing.

Suspense always worked Nelson into a state of almost hysterical excitement. With the feverish energy born of that state of mind he had not only caught up his enemy but passed him. He immediately put to sea again and started in chase of the fleet that was really quite close to him, and while Napoleon quietly landed his army at Alexandria, Nelson was scouring the Mediterranean, and did not sail from Syracuse on his final chase until July 24, when Napoleon had been a month in Egypt, and had already captured Cairo after fighting two great battles.

This time the chase was a short one, and on August 1, 1798, was fought the battle of the Nile.

It may be useful at this point to estimate the change that had taken place in the Mediterranean in the course of the last year. 1798 opened with the French in power in Corsica and the dependent islands, from which they had expelled the English, in Genoa, which they had annexed, and in the Ionian Islands, which they had torn from Venice. In the course of the year they captured Malta, and for a few weeks they also held Alexandria.   The English held nothing, and were supposed to be finally expelled from the waterway.

The close of 1798 saw the commencement of the blockade of Malta, the third capture of Minorca by England, which took place in November, the cooping up in Egypt of all the best generals of France, and an army of 30,000 men, and the ejection of the French from the Ionian Islands by a joint force of Russians and Turks.

But the year 1798 saw something far more momentous than even those startling events.   It saw a change of men's minds in respect to the Mediterranean.   Prior to 1798 the Mediterranean had been regarded as one of the minor European complications, which every diplomatist must needs master, but which had no very vivid or peculiar interest for England more than for any other nation.   It was one of the perpetual sores of Europe, like the questions of the Duchies, of the Low Countries, of the banks of the Rhine.   The idea that it led anywhere seems to have been hardly grasped ; and authority was on the side of England cutting herself loose altogether from an embarrassing and unprofitable connection.

Napoleon's action ought to have changed all that.   It was an object-lesson of the most startling importance.   It demonstrated that if England withdrew altogether from the Mediterranean, the withdrawal would operate in favour of France, and that if the English were ignorant of the use of the Mediterranean the French certainly were not.   Moreover the use that they made of it was so menacing to England that, in spite of the habitual anxiety of the Ministry to evacuate everything, they soon found themselves compelled to cling convulsively, if blindly, to every point of vantage that could be acquired.

Napoleon's policy always was to lull men to sleep until the moment came to strike.   This policy was most forcibly exemplified in his dash on the East.   He had contrived to expel his adversary from the Mediterranean, pin down her fleets in distant waters, complete his vast preparations at

Toulon, capture Malta, and get well on his way to Alexandria before any living Englishman had guessed his purpose. Even then it was only a stroke of divination on Nelson's part that revealed Egypt as the French objective : the Admiral's official instructions could imagine nothing more original than a descent on the coast of Ireland.

But when Napoleon made his second attempt to dominate the Mediterranean he made it in far less favourable circumstances. England was wide-awake to the danger, and for her, a commercial Power, the danger was nothing less than a menace to her existence. This second attempt began with the rupture of the Treaty of Amiens.

Five years had passed between the collapse of the first attempt in 1798, and the commencement of the second in 1803. They were years of intense anxiety for England—an anxiety nowhere more clearly manifested than in the indulgent terms granted to France by the Treaty of Amiens and the extreme reluctance with which the renewal of the war was contemplated. The immense mass of reading bearing on the history of the Mediterranean route to the East can only be followed here in outline. Some portions must be neglected—in particular the debates in Parliament. Yet one of these is intensely instructive—the debate prior to the conclusion of the Treaty of Amiens. In particular Mr. Pitt's speech must be mentioned. Mr. Pitt said—" The external trade of England is with the East Indies, the West Indies and the Mediterranean. It fortunately happens that the chief British conquests have been in the Mediterranean ; they include Egypt, Malta, Porto Ferrajo and Minorca. To give these up costs nothing ; to retain them serves but to mortify the pride of France—a dangerous course." In Mr. Pitt's eyes the Mediterranean was not a route at all. It was a *cul de sac* ending at Smyrna, simply an inland sea, where there happened to be some trifling trade-interest, £112,000 a year or so. Therefore, he concluded, let England not burn her hands for such trifles, but evacuate the Mediterranean altogether. By doing this she would avoid wounding the pride of France and would lose nothing material. In fact he was in favour of restoring the situation of 1797—a situation that sent consols down to 51.

Great attention should be bestowed on this speech, because it illustrates to perfection the position (that will be found, perhaps, a sound one) that with the solitary exception of Cyprus the course of England along the Mediterranean has been mainly a popular impulse. Blunderingly, but still with ever-increasing force, England has driven along the waterway from stronghold to stronghold. But if Cabinets had had their way the progress would have been constantly checked.

Everything now turned on Malta—the peace of the world hung on the ownership of the little island. It is commonly said that at the Peace of Amiens England resigned Minorca and retained Malta, and in a sense that is true. She did resign Minorca, just as in every other part of the world (with the exception of India) a treaty was concluded which was very favourable to France. But the treaty did not stipulate for the retention of Malta : on the contrary, it laid down the necessity of securing the independence of the island, with most elaborate and cumbrous guarantees ; and created one of those situations that embarrass every one and satisfy no one— a situation of the kind which Napoleon loved to create.

Malta had surrendered to the English on September 5, 1800, after a siege lasting two years, during which time the blockade was only run by five vessels. Napoleon was determined to obtain possession of it. "At all costs we must be masters of the Mediterranean;" "I had rather see the English in the Faubourg St. Antoine than in Malta"—these are two of his expressions. At the same time he was taking measures to remove England from the island, committing the most lawless acts in other directions for the avowed purpose of entering on negotiations with his hands full of equivalents, for he said, "We shall never retake Minorca."

But England was wary. Everywhere else she made the most ample concessions ; even in the Mediterranean she restored Elba and Minorca. As regards the latter island Nelson had said, " Minorca is ours whenever we like to take it," so apparently this considerable concession was made at little cost. But Napoleon thought nothing gained without Malta, and England was determined not to let the strong man in again. But Lampedusa, not Malta, was demanded. It was not without reason that Napoleon marvelled at the moderation of England. Here was a fortress described by Napoleon himself as the strongest place in Europe, and indispensable to any Power that would control the Mediterranean ; and what he meant to do when he had attained that control he had made plain to all, both by words and by actions that spoke louder than words. He had taken Malta himself, or rather the place had surrendered to him, and surrendered without resistance. " It was lucky," the French said wittily, " it was lucky that there was somebody inside to open the gates, otherwise we should not have got in." What took Napoleon three days to conquer had taken England two years to re-conquer. It was now in British hands, this strongest place in Europe, this gate to the East ; and England did not propose to hold it. The cumbrous provisions of the Treaty of Amiens were accepted without demur. The island was to be restored to the Knights of St. John. This is indeed remarkable moderation—moderation that would tempt a much more scrupulous antagonist than Napoleon. He presumed on it, and drove England to ask not indeed, even in the last resort, for Malta, but for Lampedusa. Lampedusa, together with a ten years' occupation of Malta as a material guarantee, was demanded in an ultimatum presented by Lord Whitworth on April 26, 1803.

Thus the rupture of the Treaty of Amiens was made to come from England ; and the measure taken in order to secure that end was the publication of Sebastiani's so-called commercial report. Sebastiani had been despatched to the Levant on a commercial errand, and his report was published in the *Moniteur* of January 3, 1803. It occupied eight columns of the official Gazette, and was nothing more nor less than a plan of campaign, including an estimate of the force—six thousand men—requisite for the re-conquest of Egypt. The official promulgation of such a report in a time of profound peace was nothing less than an outrage, and was followed three months later by an ultimatum from England—as was expected.

Nobody knew so well as Napoleon that Sebastiani's report was altogether unsound. It was perfectly certain that nothing more could be achieved in the direction of Egypt. The net result of all his operations, from the date of his own capture of Malta to the capitulation of Cairo on June 27, 1801, had been this—that instead of extending French authority through Egypt to India, he had only succeeded in drawing the English from India to Egypt and encamping them in Malta.

In 1797, with a year's start of England, the conquest of the East—or at any rate some part of it—was certainly practicable.

In 1798, with only a week's start of Nelson, it was already almost hopeless.

In 1803, with the English in Malta and the French expelled from Egypt, the conquest of China would not have been a wilder dream.

However, the report was useful as an *agent provocateur*, and enabled him to commence his second grand attack on England with the demeanour of one who has had a quarrel thrust upon him.

The world has grown so accustomed to regard Napoleon as a man engaged all his life in fighting, that the fact is some-times overlooked that for five years of his short career he did no fighting at all. These five years—from 1800 to 1805— were passed in consolidating his own authority, in obtaining, through the Peace of Amiens, an indispensable breathing-time, in making naval preparations for his grand invasion of England. During the last two years France and England were at war, although Napoleon himself had not taken the field. He was still in the north of France, intent on the English expedition. Since he could not strike at her through India he would strike at her heart ; and true to his principle of always "working out his problem in two ways," while pushing on his preparations at Boulogne he was thrusting his arm down Italy, and preparing by dominating the Peninsula to cut the Mediterranean in two and throw the English right and left.

The latter part of his scheme is the only part that concerns this subject, and in order to realize how simple this apparently grandiose operation really was, the state of Italy prior to its consolidation under the House of Savoy should be recalled. In pre-Revolution days the Peninsula was sub-divided into the republics of Venice and Genoa, the kingdom of Sardinia, the Austrian Grand Duchies, the Papal States, and one considerable state of the second class—the kingdom of Naples and Sicily.

Thus the internal affairs of Italy were already sufficiently confused when Napoleon commenced his operations in Italy. But it may be said to have been stability and unity itself in comparison with the kaleidoscopic changes that were brought about by the French invasions—the cis- and trans-Padane republics, the cis-Alpine republic, the Ligurian, Roman, and Parthenopœan republics, the duchy of Guastalla, the principality of Piombino, the kingdom of Etruria, the principality of Benevento (conferred on Talleyrand), the principality of Ponte Corvo (conferred on Bernadotte), and twelve duchies, grand military fiefs carved out of the ancient dominions of the Venetian republic. In such a state of things, when no man could say with certainty from year to year of what state he was a citizen, it is obvious that no resistance worthy of that name could be offered to a compact body of troops. So when, at the rupture of the Treaty of Amiens, Napoleon commenced his second attempt to dominate the Mediterranean, it was no idle threat when he ordered Saint-Cyr to march fifteen thousand troops through Italy, and occupy Pescara, Otranto, Brindisi, and Taranto. The order was executed with as little difficulty as if it had been an order to cross the parade-ground. The result of it was that Napoleon thrust himself in between the English in Malta and the Russians in the Ionian Isles. This occurred in the year 1803, at a time when the success of the invasion of England was not yet despaired of. The folly of the second attempt to master the Mediterranean may become apparent later, but at this date, 1803, it may be valuable to consider what was the effect in mid-course of "working out a problem in two ways."

The problem was "to ruin England"; the two ways were to conquer her on the seas, or to conquer her by land—that is, by isolating her and cutting off her commerce. The first was more than possible (as was seen in the wars of the American rebellion), if France devoted all her attention to the problem. But for that she must have peace on the Continent, and that she might have had for the asking. The other course was equally possible if pursued singly; for if England had seen her maritime supremacy unchallenged she would not have interfered in Continental affairs.

But Napoleon must needs make deadly enemies at once of England and of the whole Continent. His first plan of action was ruined at Trafalgar in the year 1805, and he instantly turned to his other plan. Ulm and Austerlitz demonstrated its feasibility; the Continent was at his feet. His first thought was of the Mediterranean. Naples was the only solid State not yet under his control, and from the palace of Schönbrunn he deposed the King of Naples in the phrase, "La dynastie de Naples a cessé de régner." The Mediterranean was thus cut in half—in theory—by a French maritime kingdom, of which Joseph Bonaparte, and afterwards Joachim Murat, was king. The unwisdom of attempting to dominate the Continent without having secured the neutrality of England now became clear. "Lose not a moment in seizing Sicily," he wrote to King Joseph; and again later, "I will never make peace without having Sicily." But Sicily was an island, and maledictions would not enable an army to capture an island without ships, and Napoleon had lost all his ships at Trafalgar.

The Emperor's action at this conjuncture of affairs appears to have been a double blunder. At the very moment when his fleet was destroyed, he doubled the coast-line of France by turning Italy into a French kingdom. Every port of Italy, being unprotected from the English assault, became an open sore in the body of the French Empire. One half of the new French kingdom was an island, and when Napoleon thrust his giant hand down the Peninsula and occupied the southern ports, saying "Check" to England in the Mediterranean, England replied by occupying Sicily, and said "Checkmate."

The circumstances of our occupation were, moreover, peculiarly exasperating to Napoleon, for the Queen of Naples was a sister of Marie Antoinette. Sure of his prey, Napoleon had abused her violently and in public. He had called her "Athaliah," he had threatened her with being obliged to beg her bread in the streets, he had vowed that he would not leave her ground enough to bury herself in. Nevertheless the queen was still throned in Palermo, amply subsidized by England and guarded by an English fleet. To attempt to bribe after menacing, and menacing in vain, is an unpromising course of action. This was, however, the course to which Napoleon was now reduced. He offered the King of Naples the Balearic Isles (which belonged to Spain) if he would abdicate in favour of Joseph Bonaparte. The offer was declined. He offered the Hanseatic towns (which were free boroughs of the old Germanic Empire); finally he offered Ragusa and Albania (which belonged to the Sultan).

This last offer was made to the British Government direct, and was met with the immediate reply that if the Emperor would add Venice England might consider the proposal. The point of this reply lies not so much in the fact that the Venetian territories added to Ragusa and Albania would make the King of Naples once more a monarch of the second class, as in the fact that Venice was the Emperor's favourite hunting-ground, so to speak, for military fiefs for his marshals, so that in asking Napoleon to add Venice, England was turning against him his own principle of robbing Peter to pay Paul. The second attempt of France to dominate the Mediterranean had two results : first, it incalculably increased

the hold of the English on the waterway, and secondly it
gave an opportunity to the great sea Power to fasten its fangs
into the French Empire.   As the English could not be bribed
or beaten off, the kingdom of Naples bled slowly to death.
Its authority was based on a large army of occupation, which
must be fed.   The roads being bad the supplies must go by
sea.   This merely meant that the English were fed at the
enemy's expense, while the French must starve or forage; in
other words, plunder, which was to rouse the whole country-
side against them.   One small expedition only got across the
Straits of Messina—a division of three thousand men under
Cavaignac, who was glad to retreat with the loss of one
thousand men.   England, on the other hand, was able to
achieve quite a considerable military success.   On the plains
of Maida Sir John Stuart inflicted a severe defeat on General
Reynier ; and our naval successes, though each was individu-
ally small, were numerous, and in the aggregate intensely
exhausting to the enemy.   Capri was captured on May 12,
1806, and held for two years.   Ponza and Ventoliene, and, in
the south of Calabria, Scilla, Reggio, and Amantea, were all
held either by English or by English and Sicilian garrisons.

Napoleon was fairly in the clutches of the great sea Power.
But he made one more tremendous struggle before he finally
succumbed.   He could not without extreme danger venture
to move a corporal's guard from one part of the Mediterranean
to another.   But he could march a mighty army through
Europe, and on the raft at Tilsit wring from the Czar Alexander the cession of the Ionian Islands and the Cattaro.
This was in the year 1807.   Ten years before, in 1797, the
incorporation of the seven islands with the French Republic
had been a most serious menace to the influence of England.
Their addition to the French Empire had no effect whatever,
except to provide an English squadron with employment,
during what might almost be called a holiday cruise.   The
islands were occupied one after another without difficulty ; at
the Great Peace Corfu alone held out.

At the present day nothing remains in English hands of

all these conquests except Malta. Minorca had been finally evacuated at the Treaty of Amiens. Egypt was restored to Turkey; Sicily to the kingdom of Naples. It was finally incorporated in the kingdom of Italy two years before England evacuated the Ionian Islands.

The Ionian Islands, which were placed under British protection at the Great Peace, made one of the thorniest questions that ever plagued a British Cabinet. The effective domination of the islands was nominally dissociated in the treaty settlement from the authority of the suzerain Power. Although under British protection, they had been formed into an independent republic. Geographically and by racial and religious sentiment they came in late years to belong to the mainland. But in the year 1815 there was no such thing as the Greek kingdom, and no one at the Congress of Vienna dreamed of such a kingdom arising, unless it was Capodistrias, who was at that time so far from anticipating the realization of his dream that he strove to erect the islands into an independent kingdom. When, however, the Greek kingdom was established, the existing situation called for some justification. The party in England in favour of retaining the islands found that justification in the fact of the half-century's occupation, and in the notorious facts that Napoleon had considered them incalculably important as a point of support for his dash on India, and that England had been compelled to occupy them in order to baffle his plans.

The party in favour of relinquishing them was largely composed of men who had no sympathy with the views just cited. They were not moved by the consideration that the islands were a port on the way to India, because they objected to India being held by their own country.

The situation was aggravated on both sides by the extremely provocative and even menacing language employed by the Ionians—language the more out of place when it is con sidered that they owed to England everything that made life worth living.

In such an imbroglio well might a responsible Minister

cry with Mercutio, " A plague o' both your houses." But a definite measure was inevitable, and the islands were ultimately evacuated and passed over to Greece, amid infinite jeremiads over the approaching downfall of the British Empire.

Thus 1863 found England, after two hundred years of activity, reduced to Malta and Gibraltar.

The question of the Ionian Islands is now almost ancient history, but the next two steps eastward, Cyprus and Alexandria, introduce the student to the actualities of political controversy. A new Mediterranean policy has to be considered—French in its origin like the first. In contradistinction to the earlier policy it moved with great rapidity, but, like the first, it redounded ultimately to the profit of England.

To sum up, the conception of the Mediterranean as a route to the East was French in its origin. It was an idea of Colbert's. Colbert's whole policy was to develop the French navy, and strengthen France by expanding abroad, and especially by expanding eastwards. But Colbert's master did not agree with his Minister. Louis XIV. not only turned aside from this policy in favour of that of dominating Europe by military force, but he initiated England into the line that he himself abandoned, and brought about the marriage of Charles II. which gave England Bombay and Tangier—the germs of Eastern empire and the Mediterranean route to that empire.

But Charles did not value the place, for this reason: he considered it merely as a nursery for a standing army by whose aid he might subvert the liberties of England. This disreputable intrigue, joined perhaps to a real lack of experience in adventures with half-civilized people, led to the evacuation of the place—the solitary English colony on the southern shore of the Mediterranean—after an occupation lasting twenty-two years. The place had long been valued for quite other reasons than that it was on the road to the East, and when England captured the key of the Mediterranean in the year 1704, British statesmen appear to have reflected less upon the glory of the achievement than upon the additional embarrassment that the possession of the port would add to their labours. British Ministers are not to be blamed for the fact that England still holds Gibraltar, for they made six resolute attempts to get rid of the place. They were only baffled by the determination of the people that the Rock should not be abandoned. In this the people may have been right, or they may have been wrong. Their feeling on the subject may have been what Lord Townshend called it, "a violent and superstitious zeal," or it may have been a sound political instinct. The fact remains that from the first Gibraltar and all that sprang from our hold of Gibraltar was essentially a popular question and nothing else. The key of the Mediterranean remained English because no Cabinet ventured to approach the House of Commons with the proposal that it should be handed over to any other nation.

British progress along the Mediterranean was for a long time merely experimental—and not very intelligently experimental; it was almost fumbling. For more than three-quarters of a century England oscillated between Gibraltar and Minorca, holding both places with a hesitating grasp that was a standing invitation to her foes. But behind the Cabinet was the country; and if the country was stubborn about Gibraltar it was ferocious and positively bloodthirsty about Minorca. Nevertheless there was no light of counsel in that dogged resolve. In 1795 this was clear, owing partly to our previous successes, partly to a series of whimsical social incidents, and partly to the fact that the Admiral in command in the Mediterranean—Hood—was a great sailor, and had a yet greater sailor—Nelson—under his command. Owing to this curious conjuncture of incidents and accidents, for a whole year England held the fate of Europe in her hands. She remained supine at a juncture when indolence compromised the cause of liberty. No doubt Hotham was unfitted for a position of responsibility. But the point for attention is, that if either the Cabinet or the country had known what was the use of the Mediterranean, Hotham would not have been permitted to dawdle away the golden hours. The fact is that up to the year 1797 the history of the route is simply a long struggle between the Cabinet and the country : the Cabinet seeking always to retreat and the country to push on, neither knowing what it was doing. With the year 1798 the long struggle ends, not because either party found its wits, but because Napoleon showed them both where the route led. The people laid the lesson to heart, but not the Cabinet, as was clear from Pitt's speech in the debate prior to the signature of the Treaty of Amiens. With 1798 also ends the long century of sleepy Mediterranean politics. For ten fiery years there were pooled in the Mediterranean the fate of England, of France, of Italy, and of the East—in fact the fate of the world. This desperate tangle falls into shape round three clues : the three attempts of Napoleon to dominate the Mediterranean. All three failed. The first attempt ended in drawing England from India to Egypt and encamping her at Malta. The second attempt had two phases : firstly, the attempt to conquer England by sea which ended at Trafalgar, and secondly, the attempt to conquer England by the land. The second phase commenced with St. Cyr's occupation of Brindisi, was pursued at Ulm and Austerlitz, and culminated in the Decrees of Berlin and Milan ; in the Mediterranean its only result was to draw England into Sicily. The Berlin Decrees led to Tilsit, and at Tilsit the Emperor wrung the Ionian Islands from the Czar Alexander. But whereas the first attempt had been of serious danger to England, and the second attempt had been not less serious up to the date of Trafalgar, the third attempt merely sufficed to provide a small British squadron with the opportunity of seeing active service.

The Great Peace found England in the position so often coveted by France : the Mediterranean was more than potentially an English lake. England did not abuse that position, but contented herself with Gibraltar and Malta, and the *damnosa hereditas* of the Ionian Islands. When the question of the relinquishment of the Ionian Islands became actual, the Ministerial attitude of the eighteenth century had to a certain extent become a popular attitude, and the islands were relinquished amid the general applause of a party consistently opposed to British presence in the Mediterranean.

This history commenced with the statement that the Mediterranean Question was, in effect, the Eastern Question, and that a great contrast would be found between the mutual relations of Christendom and Islam in these regions at the opening of the eighteenth century, and at the opening of the twentieth century. It was not intended to imply, and it is not here implied, that the movement was in any sense a religious one. On the contrary, although England entered the Mediterranean as the heir of a crusading Power, no steps whatever were taken either to encourage Christianity or to discourage Islam. From Tangier to Khartum the movements of Christian nations have been everywhere destitute of religious significance. Islam gave way merely because when the age of invention opened she had not sufficient sense of the movements taking place around her to master the new weapons of warfare : not because her courage had declined since the days when she threatened Vienna and fairly worried England out of Tangier ; still less because the religious zeal of Christendom unchained forces that Islam could not resist.

From the map of the Mediterranean shores in 1815 it even appears as if Islam had suffered but little. The long coast of Muhammadan territory extends unbroken from Tangier to Alexandria. The memories of the invasions of Egypt and Syria are but mere memories. French and English have come and gone, and the Crescent still waves over dominions of a very wide extent. In the north, beyond the Bosphorus, there is, it is true, an ominous shrinkage. A new Power, Asiatic but Christian, unique of its kind, the Empire of Russia, is moving, and not slowly, towards the Mediterranean. But in the coast-line there is no change.

In effect during this century and a half, from 1660 to 1815, the two rival Powers, France and England, whether from

F

weakness or from prudence, left the coasts of Africa alone. Neither England, confused if determined, nor France, clear-sighted if irresolute, cared to stir those nests of hornets, the Barbary States. Their prudence is to be commended; for when it came to blows it required the combination of both Powers to crush Turkey, and even the subjugation of Algiers required very considerable efforts. The alliance of France and England was not to be thought of during the eighteenth century, and as long as the resources of Islam and Christianity remained practically the same as regards weapons of warfare, even an expedition like O'Donnell's Morocco campaign of 1860 would have been an impossibility.

The nineteenth century opened : and with it the age of invention. It soon became apparent to what an alarming inferiority those nations were henceforth condemned who did not move with their times. No nation moved faster than England; and yet her policy in the Mediterranean continued to be as perverse as it had been from the commencement. She was not elated at her position ; she made no attempt to consolidate it : she allowed the man who for nine years held an unparalleled position of strength and influence at Malta and Corfu to be succeeded by ordinary adminis-trators. In a post that called for a knowledge of the mysterious and menacing politics of Italy, of the susceptibili-ties of the stately and suspicious government of Turkey, of the thorny questions of the Barbary States, England was contented to place a series of noblemen and gentlemen, excellent soldiers some of them, but for the most part men whose administrative talents were most unequal to the demands made upon them. Her position in the Mediterra-nean, resting as it did on the commanding fortress of Corfu and the rock (impregnable at the time) of Malta, was admirably suited for offensive operations. She held the command of the sea, and saw around her not only no single nation that could claim to be her rival, but no possible combination of nations that could hope to face her success-fully. It may be safely affirmed that no other nation would have sat idly in so great a position. As for England, she allowed herself to be worried out of the Ionian Islands, as formerly she had allowed herself to be worried out of Tangier. She watched with indifference another great nation, her old rival, seize the opportunities that she herself neglected.

Just as in India it was France who first put into practice the principles upon which the Indian Empire was to be built up ; just as in the Mediterranean itself it was France who discerned what England was apparently unable to grasp, viz. that the Mediterranean led to the East ; even so when the age of invention began, it was France who first demonstrated that the Barbary States were destined to be the prey of Europe. Algiers was the first to go—a monstrous cantle cut out of Barbary. Spain followed the lead of France. The campaign of 1860 in Morocco had many sources. It sprang from O'Donnell's personal ambition as a soldier, and from his genuine patriotism as a Spaniard, which led him to hope that a successful campaign abroad would compose the jarring factions of his native land. It sprang from his need as a politician (he was at that epoch Prime Minister) to conciliate the Church, which had been grievously offended by his Church Lands Decree, and which he trusted would forgive the leader of the new crusade. There is little doubt that he looked forward to making Morocco the Algiers of Spain. England grew uneasy. At the time when she held un-challenged the mastery of the Mediterranean she did nothing with it. Now. it appeared that all her labour had been in vain, that she had struggled for a century and a half to gain points of vantage, only to see France and Spain outwit her by boldly annexing the coast of Africa, while she timidly held on to the little islets which had cost her so much blood and treasure. The plan of annexing Morocco was post-poned, England declaring that such a step would be incon-sistent with the safety of Gibraltar. But her position con-tinued to grow feebler. The cry for the evacuation of the Ionian Islands grew louder. Finally they were evacuated. It was commonly supposed that this step was but a prelude to

adding Malta to the nascent Italian kingdom and restoring
Gibraltar to Spain. As the influence of England waned, that
of France waxed greater. Concurrently with the annexation
of Algiers, she had begun to acquire indirect influence in the
affairs of Egypt. The Suez Canal was constructed, and the
influence of France rose to its highest point. It became
clear that the struggles of the seventeenth and eighteenth
centuries had not touched the core of the matter. Clearly
the road eastward was not to be settled by the ownership of
the rocky islands that had been so ardently disputed by the
great Mediterranean rivals. Barbary lay prone to the assault
of Europe, and by boldly grasping the coasts of Africa,
France had in two generations undone the work on which
England had spent a century and a half of effort. No more
propitious moment for pushing on eastwards could have
presented itself for France. A generation ago England
seemed bent upon despoiling herself of the empire. Russia
was tied by the Treaty of Paris. Her two great rivals were
therefore out of the way. She had, with the assistance of
England, thrown back the approaching Russians, and England
herself, after languidly and irresolutely deliberating for the
period of a generation whether or not it was worth her while
to remain a Mediterranean Power, had been completely outstripped by her more energetic neighbours. There seemed
to be no reason why the Mediterranean should not become a
French lake.

But the fate of Louis XIV. and Napoleon I. overtook
Napoleon III. ; the evil genius of France interposed and
robbed her of the fruits of her statesmanship. That extreme
jealousy of rivalry on the Continent of Europe, which had
twice before been the ruin of France, produced the disaster
of 1870.

Had there been any strong national feeling on the subject,
the French Government would not have dared to risk the loss
of the Mediterranean for the sake of aggrandizement on the
Continent of Europe. But however satisfactory this explanation of the failure of France may be to the essayist, it is not

to be expected that France will be equally philosophical ; nor can one avoid feeling the profoundest sympathy with her. One year changed everything. The downfall of France was the signal for Russia to tear up the Treaty of Paris and resume her march on the Mediterranean. England once more showed signs of life. An Imperial statesman—the first since Chatham—was called to the Premiership. What was then done may have been right or wrong, wise or unwise ; but at least it was activity and not indolence.

In the wars of Turkey Islam suffered heavily. Province after province was torn away from her. Austria took Bosnia and Herzegovina ; Roumania, Servia and Bulgaria went the way of Greece; England acquired Cyprus, France took Tunis. These events had been preceded by the proclamation of the Queen as Empress of India. The generation between 1868 and 1901 decided the fate of France in the Mediterranean. In the year 1868 she was reputed the greatest military Power on the Continent of Europe : she might challenge England at sea. She had chained down Russia, completely outstripped England, and had embarked on a career of conquest and " influence " to which it was impossible to set a limit. In 1901 she is not the first military Power, and she could not dream of facing England at sea.

The acquisition of Tunis can hardly compensate her for the loss of Egypt. Tunis itself will probably go the way of Algiers—and that is not a good way. France still remains the second Mediterranean Power ; but she counts as a Power that can still damage England rather than as one likely to herself attain to a dominating position. Whereas in 1868 men asked, and asked reasonably, what limits can be set to the reach of France, it is now of England that the same question is asked. The answer is not far to seek : the limits are set firstly by the power and will of France to revenge herself, and secondly by the attitude of Russia when she becomes a Mediterranean Power.

It now becomes possible to trace in some detail the history of a few of the events alluded to in the course of the preceding six pages. First in order of time and also importance comes the French conquest of Algiers. Up to the year 1815 the history of England in the Mediterranean proceeds on definite lines. The rivalry between France and England is expressed by periodical struggles for certain points of vantage, while the intervening years are comparatively uneventful. The actual collisions between the two Powers are events of some historical importance, and from the naval and military point of view of considerable importance. But they are chiefly memorable because they form dramatic moments in an otherwise uninteresting story. Such moments are the battle of the Nile, the great siege of Gibraltar, the hardly less noble defence of Minorca by Murray, the dramatic arrest and execution of Byng. But when the smoke of battle rolls away, the intrinsic insignificance of these places becomes for the first time apparent.

After the year 1815 the course of history is completely changed. Instead of a monotonous story varied by a few exciting episodes, there is but a story in which " episodes " (which in this connection implies collisions between France and England) are entirely wanting. For eighty-seven years the two nations have been at peace. But although the outward current of affairs in the Mediterranean moves tranquilly, it is not to be supposed that France has accepted the situation of 1815. She has, it is true, renounced the struggle for barren rocks which characterized the wars of the eighteenth century. She no longer objects to the presence of England in Malta, or attempts to expel her from Gibraltar. Her attitude in respect of these possessions is rather that of the ground landlord towards an undesirable tenant, whose long lease is steadily drawing to an end, and whom it is not worth while to disturb during the remainder of his term, provided that the rest of the estate is not injured by his presence.

The rest of the estate, to carry on the metaphor, is the coast of Africa. It is here that France during all this time has been persistently working out the position that she appeared to have abandoned. During the fifty years that followed 1815 she moved steadily on to the position which she occupied in the year 1868. Her progress is marked by the gradual subjugation of an extensive province and the development of a consistent Eastern policy. Of the Mediterranean events of the hundred and fifty years that followed the entry of England on the scene, something like a connected narrative in a short space may be drawn up. Such a narrative may move swiftly if it neglect military details, and may nevertheless give an adequate review of the history of the time. But after 1815 it must be a more deliberate narrative. In however great detail the course of events be dealt with, there will remain much that must be uncertain, much that is unexplained, and which will probably remain so until the seals are lifted from the official records of that period.

The French assault on Islam was made, not in Morocco, where England made her first appearance in the Mediterranean, but in the neighbouring province of Algiers. This may have been for the reason that Spain, although inactive, had come to look on Morocco as her destined prey. Or it may have been for the reason that France had sent several expeditions to that part of the African coast in earlier years. These expeditions had had fortunate results ; but no event had so completely demonstrated the helplessness of Islam in the face of modern discipline and modern arms of precision as that which had been led against Algiers by Admiral Lord Exmouth in 1816. Throughout the seventeenth century the cities of Algiers and Tunis had been renowned for their wealth and strength. Their wealth was considerable, although greatly exaggerated. Their strength was undoubted, considering the nature of the armaments likely to be brought against them. Two punitive expeditions were despatched against Algiers by Louis XIV.— the first under the command of the eccentric Duc de Beaufort in the year 1664, the second under Duquesne in the year 1681. In the latter expedition bomb-ketches were for the first time employed against regular fortifications. These were succeeded by the expedition of 1684, which resulted in the surrender of eleven hundred slaves by Algiers and Tripoli.

Tunis did not await the assault, but came to terms on August 30, 1685. Treaties were signed : that with Algiers being dated April 25, 1684, and that with Tripoli June 25, 1685. The medal struck in commemoration of these events, and bearing the inscription, " Affrica supplex, confecto bello piratico," appears to claim too much. The more considerable exploit of Lord Exmouth, one hundred and thirty-two years later, left Africa very far from being subjugated : while the Admiralty was so little disposed to look on the expedition as a war, that only after prolonged hesitation was a medal granted to the forces employed in the bombardment.

Nevertheless though it received but little attention at the time, Lord Exmouth's expedition must be narrated here, if for no other reason than that it revealed the weakness of Algiers in the face of a determined enemy.

The command in the Mediterranean in the year 1815 was held by Admiral Sir Edward Pellew, G.C.B., created (at what was supposed to be the end of the war) Baron Exmouth of Canonteign, and subsequently promoted to a viscountcy for his services at Algiers. On Napoleon's escape from Elba Lord Exmouth was ordered back to his command. The marines of the British fleet held the castle of St. Elmo from May 20 to May 23, 1815, during the interregnum following on the battle of Tolentino and preceding the Austrian occupation of Naples. Besides this considerable service to the cause of order, a service which was rewarded by the Grand Cross of St. Ferdinand, Lord Exmouth was able to save the city of Marseilles from serious commotion by again landing the marines and some blue-jackets to co-operate with the garrison of Genoa in repulsing Marshal Brune. But no serious naval operations were undertaken ; nor were any possible where there was no hostile fleet in existence. The ships went into winter quarters at Leghorn, and awaited instructions. When these instructions reached the commander-in-chief they were found to include the duty of sounding the states (one can hardly call them powers) of the

Mediterranean as to their willingness to confide their interests to the British flag. Naples accepted : the Holy See declined. Lord Exmouth appears to have made the safe assumption that the other states would not decline to receive back their liberated slaves from his hands.

It must be left to a naval expert to say whether the errors in the Admiralty charts of Algiers of this date were exceptional or not. To a civilian they appear to be considerable. They include the statement that the sea-front of the city of Algiers was four miles in length instead of one. But whether exceptional or not, they would appear to be sound evidence that up to the year 1816 Christendom had preferred to contemplate the defences of Algiers from a respectful distance. In fact a pirate city that mounted one thousand pieces of artillery on its walls was in a position to command respect. On January 25, 1816, Captain Warde of H.M.S. *Banterer* was ordered to proceed to Algiers as a spy. A pretext was to be assigned for his presence ; although not a very sound pretext, had the Consul been inquisitive. Captain Warde was directed to state that his work was connected with the new constitution of the Ionian Islands then being drafted by Sir Thomas Maitland, with whom Lord Exmouth was to be joined. Ridiculous though this pretext was, it served. Warde proved himself to be a discreet and dexterous spy. He made ample observations and drew a plan of the sea-front and defences. Lord Exmouth was now confident that if negotiations failed it would be possible for him to assign positions to his attacking ships with the certainty that they would be able to make their fire effectual. On March 21, 1816, he issued the following General Memorandum from on board H.M.S. *Boyne* in Port Mahon—

" The Commander-in-Chief embraces the earliest moment in which he could inform the fleet of his destination, without inconvenience to the public service.

" He has been instructed and directed by His Royal Highness the Prince Regent to proceed with the fleet to Algiers, and then make certain arrangements for diminishing at least the piratical excursions of the Barbary States, by which thousands of our fellow-creatures, innocently following their commercial pursuits, have been dragged into the most wretched and revolting state of slavery.

" The Commander-in-Chief is confident that this outrageous system of piracy and slavery rouses in common the same spirit of indignation which he himself feels ; and should the Government of Algiers refuse the reasonable demands he bears from the Prince Regent, he doubts not but the flag will be honourably and zealously supported by every officer and man under his command in his endeavours to procure the acceptation of them by force ; and if force must be resorted to, we have the consolation of knowing that we fight in the sacred cause of humanity and cannot fail of success.

" These arrangements being made at Algiers and Tunis, the Commander-in-Chief announces with pleasure that he is ordered to proceed with all the ships not on the peace establishment to Spithead without delay; except the *Bombay* bearing the flag of Rear-Admiral Sir Charles Penrose, which ship is to be relieved by the *Albion*, daily expected.

(Signed)   " EXMOUTH.

" N.B.—This General Memorandum to be entered in the public order book and communicated to the respective officers, seamen and marines of the fleet."

There is some verbiage here: customary and perhaps useful verbiage. But the memorandum is set forth *in extenso* in order to make it clear that no measure of annexation was in contemplation. The expedition was only conditionally punitive ; and it was to be the last service—if it should amount to active service—exacted before the ships were paid off. It is to be observed that Lord Exmouth was not instructed to bring about the abolition of slavery. He was only entitled to say that it would be agreeable to the Prince Regent if slavery could be abolished.

At Algiers the Ionians, who had recently become what the punctilious language of diplomacy called " Citizens of the Independent Septinsular Republic," under the protection of Great Britain, were released without hesitation. The slaves from Naples were bought out at £100 a head, and those of Sardinia at £60 a head. At Tunis and Tripoli the Governments consented to take the Prince Regent's invitation into favourable consideration.

Prior to Lord Exmouth's expedition the navy of the United States had undertaken a punitive expedition to Algiers on its own account, had fought a successful action with a division of the Algerine fleet (June 17, 1815), and had obtained (June 30, 1815) from the Dey concessions which justified Commodore Decatur in stating that the United States were placed on higher ground than any other nation. While visiting Tunis and Tripoli, Lord Exmouth received orders to return and arrange that England should receive most-favoured-nation treatment at Algiers. He urged the question of slavery anew, but was unable to procure more from the Dey than the promise that he would send a plenipotentiary to Constantinople to sound the Imperial Government on the question. Supposing the Porte to be favourable to the Prince Regent's views, the instructions furnished by the Dey to his envoy would contain (so it was promised) authority to proceed to London and enter into negotiations on the subject. It is necessary to record that this scanty concession had been preceded by the arrest of the British Consul, by gross impertinence to Lord Exmouth, accompanied by personal violence, and something like a threat to murder his staff. On June 23, 1816, Lord Exmouth landed in England. Exactly one month earlier the armed forces of the Dey had committed an outrage which could not be overlooked. On Ascension Day, May 23, a body of coral-fishers were massacred at Bona, two hundred miles to the eastward of Algiers. Lord Exmouth was at Algiers at the time of the incident ; but news travelled slowly ; he had weighed anchor before it reached Algiers, and only heard of it on his arrival in England.

He had landed in much uncertainty of mind. He feared that his attempts to put an end to slavery would be unpopular in England. It was often maintained that England suffered little or nothing at all from the Barbary corsairs, that other nations, less powerful at sea, suffered a great deal, and that, as a consequence, the pirate cities were a source of considerable indirect profit to the Mediterranean trade. This is not a lofty attitude of mind, but there had been similar examples of it in Mediterranean waters. The smuggling trade between Sicily and Malta, while the latter island was under quarantine, had been profitable. The British merchants in Sicily were even in a position to persuade the Sicilian Government to refuse *pratique* to Malta long after the plague had subsided, so that illicit gain might still be made by the Sicily traders. All English trade was not indeed so unscrupulous. Lord Exmouth feared what he called "the old mercantile interest," and looked forward to finding himself in an "awkward situation" in consequence of his excess of zeal at Algiers.

But the anti-slavery agitation was strong. Parliamentary action, so far as it had gone, had been in the direction of disapproving the agreement to ransom slaves. The sums actually expended for that purpose had been already criticized.

At the height of the discussion came the news of the massacre at Bona. It was now no longer a question of inconvenience, or incivility to be endured from the Barbary States. It was no longer a question of how much could be suffered for the sake of a quiet life. The massacred fishermen were not indeed Englishmen, but any nation might be victimized. England was clearly the only Power capable of dealing with the situation. Lord Exmouth's fleet was largely reinforced, and shortly set sail.

As this was the last occasion before 1882 when England interfered with an armed expedition on the African coast of the Mediterranean, it may be useful to indicate what was her attitude of mind towards African questions. In comparison with the amount of provocation that drew from France the huge expedition of the Comte de Bourmont, it must be admitted that England showed great patience.

Nothing would induce England to interfere unless she was positively compelled to do so : annexation she did not dream of.

Allusion has already been made to the gross impertinence of arresting, and arresting with ignominious incidents, the British Consul and two officers of Lord Exmouth's staff on the occasion of his first visit to Algiers. The affront was ignored.

Earlier in the year there had been complications with Tripoli. Two English vessels had been seized and confiscated. The incident was dealt with as follows :—

THE GOVERNOR OF MALTA TO THE PACHA OF TRIPOLI.

"*March* 16, 1815.

"SIR,

"It is with extreme regret that I feel myself under the necessity of addressing your Highness on a subject that may, in its consequences, compromise that amicable understanding and friendship which has long subsisted between your Highness and the British nation. But it is impossible for me to admit of any open indignity to be shown to the British nation or the British flag in this neighbourhood without immediately taking such notice of it as is suitable to the dignity, power and maritime supremacy of the king my master.

"I therefore have the honour to inform you that I have sent instructions to His Majesty's Consul, Colonel Warrington, to demand instant redress and reparation for the insult offered to the British Crown in permitting two vessels with British colours flying to be seized in the port of your Highness, and under the guns of your works ; and I have further directed the Consul to intimate to you that he can enter into no communication of any kind till such redress be given."

This closed the incident. But clearly, if England had contemplated annexation the matter might have been so handled as to justify hostile action. The Algiers expedition was despatched, did its work and returned, and in the next year an even more outrageous breach of the peace brought about such relations with Tunis as would have served as an excellent pretext for annexation. This was nothing less than a buccaneering expedition to the narrow seas resulting in the capture of two wealthy prizes, which might have been carried off to Tunis but for the amazing assurance of the privateers, who conducted their operations under the actual observation of a king's ship. France had to deal with no such difficult incident as this in the days before her conquest of Algiers.

It was an extremely difficult matter to handle. But the difficulty was rather with the Consul than with the Bey of Tunis. As regards all these half-barbarous states the policy of England was one of contemptuous toleration. They were not civilized states, but they had to be treated as such. The sole aim of England was therefore to keep things as quiet as possible, and not for the first time the difficulty came from a British officer. His maladroit handling of the Bey was not allowed to lead to serious difficulties. The question on this occasion was not how to obtain sufficient redress for this insult, but how to reduce the measure of this barbarian's offence, so as to make it practicable to overlook it as far as might be, and to make the inevitable compensation as little offensive as possible to his dignity. It is unnecessary to say that England did not go to war with Tunis over her buccaneering. The incident was minimized.

Here then are three examples, one from each Regency, of the way in which England dealt with Barbary questions. Nothing less than the massacre of Bona was taken seriously. The consequences of the massacre of Bona were serious.

Six days after Lord Exmouth's return to England after his first visit to Algiers, the Admiral was sounded as to his willingness to assume the command of a punitive expedition. He at once accepted. On July 25, 1816, he hoisted his flag on the *Queen Charlotte*, left Plymouth on the 28th, and made Gibraltar August 9. Here he met with a Dutch squadron, whose Admiral offered to support him. The Dutchman's assistance was accepted. At daybreak on August 27 the combined fleets were in sight of Algiers. On the way from Gibraltar to Algiers Lord Exmouth had fallen in with the *Prometheus*, bringing the news that the British Consul was in irons, and that the surgeon, three midshipmen and fourteen seamen of the *Prometheus* were kept prisoners in Algiers.

At two o'clock in the afternoon of August 27, 1816, the Dey's batteries opened fire. His Highness had previously rejected Lord Exmouth's ultimatum, and the British fleet had drawn into the harbour until within range of the Algerine guns. The ships engaged in the attack were the *Queen Charlotte* (108, flagship), *Impregnable* (104), Admiral Milne; three 74's, the *Superb*, the *Minden*, and the *Albion*; the *Leander* (50), the *Severn* and the *Glasgow* (40's), the *Hebrus* and the *Granicus* (36's), the *Mutine* and the *Prometheus* (16's); four bomb-ketches, the *Infernal*, the *Hecla*, the *Fury*, the *Beelzebub;* the *Cordelia*, the *Britomart* (10), and the *Express* schooner. In addition England was so fortunate as to have the help of the Dutch squadron under Admiral Von de Capellen, consisting of five 36's, viz. the *Melampus* (flagship), the *Frederica*, the *Dageraad*, the *Diana*, and the *Amstel*, with the *Eendragt* (24).

To oppose this considerable armament the Dey had called in his regular fleet, which only numbered, however, four frigates, five corvettes, and thirty-seven gun-boats. These figures are valuable; they indicate the strength and the weakness of the pirate city. Algiers was not a maritime Power. It was a predatory Power with numerous boats of sufficient size to make themselves dangerous to unarmed craft, and resting upon the basis of a strong land fortress. That fortress was in fact impregnable, except to the assault of one of two navies—those of France or England. On the present occasion an army of forty thousand men was in readiness to repel a land attack, should any be attempted. It was decided, however, that the destruction of the seaward defences would probably bring the Dey to reason. The ships took up their positions in accordance with the plan of the city drawn by Captain Warde ; and at two in the afternoon the battle began. The engagement became general at a quarter to three and lasted until nine at night. The last gun was fired at half-past eleven, and at two o'clock in the morning the fleet was once more at anchor out of range of the few guns not dismounted in the course of the action. The British loss was 883 killed and wounded, that of the Dutch 65. The losses of the Algerines were not known precisely. They were variously estimated at from 4000 to 8000 killed and wounded. But the figures are unimportant; the punishment was sufficiently severe for the Dey to hesitate before refusing a second ultimatum. As the Admiral evidently did not intend to land an army, and so give a chance of striking a return blow, a renewal of the conflict could only mean that the fleet, having destroyed the ships and batteries of Algiers, would next proceed to lay the town in ruins. There have been barbarian leaders of sufficient fortitude to slay the British Consul and the other prisoners, withdraw to the desert, and defy the fleet to do its worst. Fortunately the Dey was not a man of that stamp. He received the second ultimatum, identical with the first, at daybreak on August 28, and agreed to submit.

In effect, unless he was prepared to retreat to the desert, there was no other course open to him. In the face of Christendom armed, Islam, it was clear, was helpless. Of his gun-boats, manned by desperately brave men, thirty-three out of thirty-seven were sunk before they came within striking distance of the enemy. This was a greater proportional slaughter than befell at Omdurman. It is true that the British had also suffered severely. Lord Exmouth himself was struck in three places ; the *Impregnable* had fifty men killed, and every ship of the squadron had been mauled. Still the fleet could fight again, and the Algerines could not. The following terms, therefore, were accepted on August 30, 1816 :—

1. The abolition of Christian slavery for ever.
2. The surrender of all Christians actually in slavery.
3. The repayment of money paid as ransom since January 1816.
4. Reparation to the Consul.
5. A public apology to the Consul.

The last condition was probably the only one of the five that cost the Dey anything to fulfil. Undoubtedly the necessity of apologizing in the presence of his own Ministers must have been bitter, and may even have endangered his tenure of the throne. In other respects the terms of peace were easy. The total number of slaves liberated by Lord Exmouth during his two visits was 3003. Their nationalities are significant and are here set forth in full :—

AT ALGIERS.

Neapolitans and Sicilians	1110
Sardinians and Genoese	62
Piedmontese	6
Romans	174
Tuscans	6
Spaniards	226
Portuguese	1
Greeks	7
Dutch	28
English	18
French	2
Austrians	2
	1642

AT TUNIS.

Neapolitans and Sicilians	524
Sardinians and Genoese	257
	781

AT TRIPOLI.

Neapolitans and Sicilians	422
Sardinians and Genoese	144
Romans	4
Hamburghers	4
	574

There is no nation that does not extol its own exploits as unselfish and chivalrous. It is quite customary to explain, on entering on a campaign, that other nations will benefit by the results much more than the belligerents. It remains

G

therefore merely to point out that only eighteen slaves
out of 3000 were English, and that English trade positively suffered by the abolition of piracy.   An incident of an
earlier punitive expedition puts Lord Exmouth's work into
an amiable light.   It is on record that in the year 1684
Damfreville sent all the English slaves back to captivity,
because they maintained to his face that it was the terror of
the name of England that had procured their release.   This
was ungrateful on the part of the English ; it deserved to be
noted in Damfreville's despatch ; it deserved to be mentioned
to the British Ambassador.   But a few graceful words would
have sufficed to make up for the boorishness of some rough
sailors.   It is pleasing to recall that Lord Exmouth's noble
work was marred by no such exaggerated assertion of his
master's greatness.   The Christian slaves of all nations were
liberated without discriminating between those of the states
who had welcomed him as a champion, and those of the
states who had declined his assistance.   It is also well to
remember that at this date (1816) England was mistress of
the Mediterranean in a sense that neither she nor any other
nation has been since Pompey destroyed the pirates.   In
1816 she could have faced the world in arms without greater
strain than she now endures in order to face two other
Powers successfully.   Had England been tempted by the
idea of annexation, the temptation would have been irresistible.   It does not appear that aggressive measures beyond
the bombardment were designed by the Ministry.   When
immediately after that event a piratical expedition to the
narrow seas was organized, far from using the outrage as an
excuse for annexation, or for a second punitive expedition,
the incident was minimized.

The attitude of France towards Algiers was different.
French historians have condemned Lord Exmouth's expedition as sterile.   In a sense that is true.   The bombardment of Algiers did not deter the Algerines from continuing
their piracies.   Probably it would have needed one such
expedition every ten years to keep Algiers in order, if France

had not annexed the country. The peace of the Mediterranean gained decidedly by the action of France. It is no more open to a candid historian to censure the action of France in Algiers than it is to censure the action of England in Egypt.

Nevertheless, and in spite of all disclaimers, it is evident that France contemplated the annexation of Algiers. France in fact inaugurated the policy of moving eastward by the process of seizing on the northern coast of Africa, just as England in earlier days chose the plan of moving eastward by a series of stepping-stones. Similarly, just as the action of England roused the jealousy of France, even so the action of France roused the jealousy of England. But there is a difference of degree in the depth of resistance offered by each of the two countries to the progress of the other. France, both in her own active progress, and in her resistance to the progress of her neighbour, was far more acrimonious than England. Throughout the eighteenth century she was incessantly occupied with the endeavour to beat England back. She has never acquiesced in the presence of England or regarded the English as other than intruders in the Mediterranean. So soon as it became clear that England was not to be easily expelled, she turned to the alternative policy of progress eastward by land.

It was now the turn of England to become anxious and inquisitive. But after an exchange of notes, which will shortly be set forth at length, she ceased to remonstrate. Already in 1841, Lord Aberdeen could write of "ten years of acquiescence" in the French occupation, and refer to such acquiescence as being "entirely consistent with propriety and duty." From whatever motives England so early gave up the suspicious attitude of 1830, it must be admitted that her bearing compares favourably with that of France towards the not dissimilar action of England in Egypt. Nineteen years have passed since the latter occupation became effective. It would be waste of time to recount the thousand hindrances which England has had to overcome in the course of these nineteen years: hindrances which it was at any moment in the power of France to remove, had she accepted the situation.

In returning to the earlier years of the century, it seemed as if the new policy of France was about to neutralize, if not nullify, the English policy of stepping-stones.

The final rupture between France and Algiers was the result of a long series of wrangles arising out of some "concessions" long enjoyed by the French on the Barbary coast. A temporary settlement of outstanding difficulties was arrived at on July 24, 1820. In consequence of an unsuccessful interference by Sir Harry Neale on behalf of Consul Macdonnell in the year 1824, the Dey assumed a more offensive attitude in affairs where he had to deal with the French. In June 1825 the French Consul's house was searched for ammunition which the Dey suspected the Consul of furnishing to the Kabyles, who were in rebellion. Acts of buccaneering became more frequent. On October 28, 1826, two French frigates were ordered to Algiers to exact reparation for some pirate work. But at the same time the Dey, far from feeling overawed by this demonstration, was urging in Paris some claims of his own against the French Government.

M. de Villèle was Prime Minister, M. de Damas Minister for Foreign Affairs. The Cabinet was divided. The "Eastern Question" was urgent: but it was supposed to centre in the affairs of Greece. The Consul was in consequence instructed to press the claims of France. But M. Deval, the Consul, naturally found himself badly received at the Court of the Dey. A long series of irritating interviews closed with an outbreak of temper on the Dey's part. He struck M. Deval at a public reception.

There was no dissension in the Cabinet when M. de Damas proposed that the Dey should be required to apologize, either personally or by his Ministers. At the moment of apologizing, the French flag was to be hoisted on the wall of Algiers and saluted with one hundred guns. This measure

of satisfaction having been refused, the French Consul withdrew, and Algiers was declared to be in a state of blockade
June 15, 1827. The unrepentant Dey pillaged and destroyed
the French store-houses at Bona and La Calle.

But a pirate state does not suffer much from a blockade.
Algiers had no credit in the money market to be injured, and
very little legitimate commerce to be destroyed. Privateering
had long ceased to be remunerative—in fact ever since Lord
Exmouth had annihilated the Algerine navy. The British
bombardment had been slighted as a sterile exhibition of
force. It was now borne in upon the French Cabinet that if
the bombardment had effected little, the blockade was effecting nothing at all. M. Deval had long been an advocate of a
land expedition, and his views were now to find official
expression.

On October 14, 1827, the Minister for War, M. de Clermont-
Tonnerre, formally advised the king to annex Algiers. Three
months earlier—on July 6—France had signed the Treaty of
London, by which the contracting Powers were bound to
abstain from fresh acquisitions at the expense of Turkey.
But M. de Clermont-Tonnerre glided over the fifth clause of
that Treaty, maintained that Algiers was only a nominal
dependency of Turkey, and justified the course that he was
advocating by the parallel of England's action in Burmah and
Russian action in Central Asia. The first Burmese war had
been concluded eighteen months earlier. The Treaty referred
to by M. de Clermont-Tonnerre was that by which the independence of Greece was secured.

This Treaty presents the Eastern Question in little. Whenever a piece of territory near to the heart of Islam is hewn off,
the high contracting Powers meet and solemnly lay down the
limits within which aggression is to be allowed to take place.
But hardly is the ink dry, than the work of aggression
goes rapidly forward on the borders of the huge, defenceless
empire.

The word "conquest," now spoken for the first time, was
followed by a reasoned plan of campaign. Sidi-Ferruch, to

the west of Algiers, was fixed on as the landing-place. Thirty-three thousand men with 150 guns was to be the land force, two millions sterling the expense, of which part, if not all, would be recouped from the treasures of the Dey. This is frank and straightforward speech, as becomes a Minister addressing his sovereign. There is no more talk of punitive expeditions, or the suppression of Christian slavery. England and Russia are profiting by the decadence of Islam, and France must not be left behind. The situation is defined.

For the moment nothing was done. M. de Clermont-Tonnerre had insisted that the expedition should be put in hand at once, so as to commence operations in April. M. de Villèle on the other hand, for political reasons, was resolved on an immediate appeal to the constituencies. The Chamber was dissolved, and M. de Villèle was driven from office.

On January 4, 1828, M. de Martignac kissed hands as M. de Villèle's successor. Foreign affairs were entrusted to the Count de la Ferronays. As compared with his predecessor, the new Minister had this advantage, or disadvantage, that he perceived that public opinion was indifferent to the Algerine question. The contrast between England and France was on this occasion as striking as ever. The English Cabinet has ever moved along the Mediterranean with reluctance, pushed on by the impatient people. The French Cabinet, with more enterprising views and a stronger administration, has too often had reason to mourn that the people was not with them. Accordingly in the House of Peers on February 15, 1828, the Foreign Minister went so far as to adopt the indulgent attitude towards the States of Barbary that England had throughout maintained.

But if M. de la Ferronays understood France he misapprehended Algiers. A first unsuccessful mission was followed by a second, the king himself adding this note—" The mission of Lieutenant Bézard to Algiers is approved ; his instructions to be conciliatory but of becoming firmness." " Becoming firmness" in dealing with a pirate state must be represented by an armed force, otherwise it is a mere phrase. The immediate result of the second mission was that the Dey formulated his demands. They included the raising of the blockade, and the repayment by France of the Dey's war expenses. The next offer came from France—" The blockade should be raised if the Dey would consent to send an officer of high rank to France, to treat of the questions at issue between France and Algiers." The Dey declined the offer. A final attempt at compromise was made by M. de la Bretonnière, who was despatched to Algiers at the end of July 1829. The Dey dismissed him with menaces, and as his ships withdrew, the flag was fired on by the batteries of the harbour. This amounted to a *casus belli*. M. de Clermont-Tonnerre had been prepared to act without one. M. de Martignac was less adventurous. But M. de Polignac was now in office and was prepared to act on the ground that " Le Dey Hussein avait outragé le pavillon français, et le pavillon parlementaire, l'honneur d'une grande nation et le droit de toutes les nations." Perhaps this language was not exaggerated, although it is but just to record that the Dey had apologized for the outrage, and had dismissed the officers who were responsible for it. But if the life of a French commander-in-chief had been threatened, as had happened to Lord Exmouth; if officers of the French army had been arrested, and marched through the streets of Algiers with their hands tied behind them, as had happened to Captains Pechell and Warde ; if the French Consul had been thrown into irons, like Consul Macdonnell; if Algerine corsairs had been caught *flagrante delicto* and sailed into Le Tréport, as the *Alert* had sailed them into Ramsgate, what stronger language could the French have employed ?

When a Government has determined on a conquest, the *casus belli* is a matter of hardly more than academic interest. France was about to annex Algiers, not because she had suffered more than England at the hands of the Algerines, but because she would not be left behind in the race for annexation. The English with greater provocation had been contented with a punitive expedition. But the French made the most of the indignities offered them by the Dey, and were prepared as early as 1827 to proceed to annexation without any excuse at all.

But again there were dissensions in the Cabinet. M. de Bourmont (afterwards Commander-in-Chief of the French army of conquest), the Minister of War, assumed that the expedition was settled, and applied himself to details in concert with the Minister of Marine. M. de Polignac was of a different mind. At this crisis Egypt was under the control of Mehemet Ali, one of those great figures whose appearance has so often changed the face of the world.

The Pacha proposed to subjugate and administer the Barbary States if France would advance the sum of £800,000 and give four ships of war. M. de Polignac appears to have welcomed the idea of getting rid of the Algerine question so easily, but to have but scantily appreciated the difficulties in the way of any such apparently simple solution of the difficulty. The consent of the Sultan must be obtained : for however slight his control might be over Algiers, it was effective, for the present at any rate, over Egypt. The Porte refused to allow the Pacha of Egypt to take part in any such enterprise. This is perfectly intelligible and quite what might have been expected. The Porte had no desire to see an already powerful vassal acquire control over so vast an extent of Africa. But M. de Polignac did not seem to think that the refusal of the Porte to grant the necessary firman was a material obstacle. Nor was it if France and England had been really resolute. But in addition to slighting the Porte M. de Polignac disputed the Pacha's terms. The loan could not be £800,000, but might be £400,000. The ships could not be given, but they might be lent, and when Mehemet Ali objected that as a Prince of Islam he could not sail against another Prince of Islam under a Christian flag, M. de Polignac seems to have treated this objection (which was the core of the whole matter) as a piece of childishness. On January 16, 1830, he officially notified the Great Powers of his plans, and the notification was received with chilly reserve. Naturally : what Mehemet Ali was proposing was something in the nature of a *coup d'état*. It was a sound and sensible policy in essence, and when completed would have been of advantage to the Powers. The Pacha of Egypt was a ruler with whom it was possible to treat on civilized lines, and the interminable Barbary question would have received a solution, sudden but satisfactory. But to inform the Great Powers that the *coup d'état* was in course of preparation, at the time when the anger of the Porte was aroused, and the terms of the negotiation still unsettled with Mehemet Ali, is to make one doubt whether M. de Polignac ever seriously contemplated carrying his plans through. Probably the Pacha arrived at some such conclusion, for on January 31, 1830, the French Cabinet determined to undertake the conquest of Algiers, but offered to make Mehemet Ali a present of £320,000 and leave him to work his will on Tunis and Tripoli. The large capacity of the ruler of Egypt resented the trifling, and his ambition turned in another direction.

But it is to be observed that M. de Polignac, by his method of handling the Pacha's proposals, succeeded in making the Pacha himself break off negotiations.

Whether or no this was the end he kept in view from the first, he is entitled to claim that his diplomacy was successful ; for it disembarrassed France of an awkward rival, and left her a free hand in the affairs of the Barbary States.

By April 1830 France had resolved on the conquest of Algiers. The impression created in England by her preparations may be gathered from the following important correspondence.

On March 5, 1830 Lord Aberdeen wrote to the British Ambassador in Paris as follows :—

"Foreign Office, March 5, 1830.

" MY LORD,

"The extensive scale of the preparations for the expedition against Algiers, and the declaration in the speech of His Most Christian Majesty upon the subject, have naturally engaged the attention of His Majesty's Government. Your Excellency is already aware of the sincere desire which His

Majesty entertains that the injuries and affronts which have been endured by the King of France from the Regency of Algiers may be duly avenged, and that His Most Christian Majesty may exact the most signal reparation from this barbarous state ; but the formidable force about to be embarked, and the intimation in the speech to which I have alluded, appear to indicate an intention of effecting the entire destruction of the Regency, rather than the infliction of chastisement. This probable change in the condition of a territory so important from its geographical position, cannot be regarded by His Majesty's Government without much interest, and it renders more explanation of the intentions of the French Government still more desirable. I have communicated these sentiments to the Duc de Laval, and have received from His Excellency the most positive assurances of the entirely disinterested views of the Cabinet of the Tuileries in the future disposal of the State of Algiers. Notwithstanding His Excellency has promised to write to his Government in order to obtain the means of making an official communication, I have thought it right to instruct you to bring the subject under the notice of M. de Polignac. It is probable that the French Minister may be desirous of affording all the explanation we can desire. The intimate union and concert existing between the two countries give us reason to expect that we shall receive the full confidence of the French Government in a matter touching the interests of both, and which in its result may be productive of the most important effects upon the commercial and political relations of the Mediterranean States.

> " I am, etc.,
> (*Signed*)   " ABERDEEN."

H.E. LORD STUART DE ROTHESAY, G.C.B.
etc. etc. etc.

At this date the Cabinet of the Tuileries had resolved upon the annexation of Algiers. Can the intention to annex a country be, in any sense, described as a " disinterested " attitude towards that country ? There will be no attempt in these pages to denounce the policy of France on the ground that her words have so little correspondence with her actions. The conditions of high policy often exact reticence, and sometimes compel tortuous dealing when an answer to an inconvenient question is demanded. Nevertheless, and as a simple matter of history, it is not to be denied that, whether unavoidable or not, the course of action pursued by France in respect of Algiers was decidedly disingenuous. Lord Stuart de Rothesay carried out his instructions, and on March 8 replied to Lord Aberdeen as follows :—

"*Paris, March* 8, 1830.

"MY LORD,

"I have been honoured with your Lordship's letter of the 5th instant, and have lost no time in communicating with the Prince de Polignac upon the subject to which it relates.

"His Excellency informs me that a communication from the Duc de Laval upon the same subject had reached him a few hours before, that he had not yet sought the king's orders, but that he should do so without loss of time, and hopes they will enable him to address a communication to that Minister containing a satisfactory answer to the questions put forward by my Government respecting the objects of the expedition, and the future destiny of the Regency of Algiers in case of success.

"He said that in the meantime he could enable me to convey to your Lordship the assurance of His Most Christian Majesty's readiness to deliberate with His Majesty and with his other allies respecting the arrangements by which the government of those countries may be hereafter settled in a manner conducive to the maintenance of the tranquillity of the Mediterranean and of all Europe.

"I have the honour, etc.,

(*Signed*) "STUART DE ROTHESAY."

THE EARL OF ABERDEEN, K.T.
etc. etc. etc.

A fortnight later Lord Aberdeen replied more precisely :—

"*Foreign Office, March* 23, 1830.

"MY LORD,

"The Duc de Laval has communicated to me, by order of his Court, the copy of a despatch which His Excellency has received in answer to the inquiry which you were instructed to make into the real views and intentions of the French Government in undertaking the expedition now preparing in the forts of France against the Regency of Algiers.

"The explanations afforded by the despatch, so far as they relate to the causes and general objects of the war, have been satisfactory to His Majesty's Government; and this satisfaction has been increased by the voluntary offer of M. de Polignac to render these explanations still more precise and clear in those points where it may be thought necessary to do so.

"His Majesty has long been sensible of the gross outrage and repeated insults which His Most Christian Majesty has sustained by the conduct of the Government of Algiers, and His Majesty has always expected that for such conduct the most signal reparation would be exacted. The additional objects which a sense of accumulated injuries has induced the French Government to give to the intended expedition are such as His Majesty cannot but approve. They are such as His Majesty has himself proposed, and for the attainment of which he has himself made considerable sacrifice.

"I am further commanded by His Majesty to express his confidence in the disinterested views of His Most Christian Majesty, and in his desire to render the consequences of this enterprise generally beneficial to the states of Christendom. It appears, however, that the character of the expedition is of no ordinary description, for if I correctly interpret the despatch of M. de Polignac, it is undertaken not so much for the purpose of obtaining reparation, or of inflicting chastisement, as of carrying into execution a project which may possibly lead to a war of extermination. Under these circumstances, the declaration of His Most Christian Majesty, that in the event of the destruction of the Algerine State he will concert with his allies the means of most effectually securing the objects proposed, can scarcely be considered as affording that entire satisfaction which we may reasonably expect to receive.

"In the development of the intentions of the French Government, as afforded by the despatch of M. de Polignac, I will not conceal from your Excellency that the entire silence respecting the rights and interest of the Porte has been observed with some surprise. It is difficult to imagine that under any change of circumstances these claims should be neglected by His Most Christian Majesty. It is true that many of the states of Europe, and France and England amongst the number, have long been accustomed to treat the Regencies as independent Powers, and have held their Governments to be responsible for their conduct; for we have neither forgotten their relation to the Porte, nor the species of sovereignty which the Sultan still exercises over them.

" It is only very recently that His Most Christian Majesty has renounced the intention of availing himself of the mediation and authority of the Turkish Government in order to effect a reconciliation with Algiers. These Barbary States are still vassal and tributary to the Porte, and when the power of the vassal ceases to exist, it is reasonable to suppose that the rights of the sovereign may meet with attention. The solicitude which His Most Christian Majesty has always shown for the preservation and welfare of the Turkish Empire forbids us to think otherwise.

" Whatever may be the means which shall be found necessary to secure the objects of the expedition, the French Government ought at least to have no difficulty in renouncing all views of territorial possession or aggrandizement. The expressions of a former despatch from the French Minister, and the substance of which was communicated by the Duc de Laval to His Majesty's Government, were sufficiently precise in this respect, and it is therefore to be presumed that the Cabinet of the Tuileries will feel no reluctance in giving an official assurance to the same effect.

" M. de Polignac is doubtless aware of the great importance of the geographical position of the Barbary States, and of the degree of influence which, in the hands of a more civilized and enlightened government, they could not fail to exercise over the commerce and maritime interests of the Mediterranean Powers. The difficulty in accomplishing any radical change in the actual state of possession, by which these interests would not be unequally and injuriously affected, is perhaps the chief reason for the existence of a lawless and piratical authority having been so long tolerated.

" Your Excellency may recall to the recollection of the French Government the conduct observed by His Majesty upon an occasion not dissimilar from the present. When His Majesty found it necessary, for the vindication of his own dignity and the reparation of his wrongs, to prepare an armament against Algiers, the instructions addressed to the commander of His Majesty's naval forces in the Mediterranean were communicated to his allies without any reserve;

and the whole plan, objects and extent of the expedition were laid open.

"Your Excellency will read this despatch to M. de Polignac, and you are authorized to deliver a copy of it to his Excellency.

"I am, etc.,

(*Signed*)   "ABERDEEN."

LORD STUART DE ROTHESAY, G.C.B.
     etc. etc. etc.

The concluding paragraph of Lord Aberdeen's despatch either refers to the resolution of the Congress of Vienna, or is an oversight, for M. de Polignac could not find the circular despatch referred to, nor, upon his writing to that effect, was he supplied with a copy of the communication referred to by Lord Aberdeen. The deliberate disavowal of the intention to annex any part of Barbary was made to Lord Stuart de Rothesay in a conversation with M. de Polignac, which he reported in the following despatch :—

"*Paris, March* 26, 1830.

"MY LORD,

       "After receiving your Lordship's despatch of the 23rd instant, I went to the Prince de Polignac, and observed to his Excellency, that although the statement of the motives for the expedition to Algiers, and the views of the French Government upon that country contained in the letter which he has caused to be communicated to your Lordship, had been received with satisfaction by my Government, that measure involves considerations upon which we are justified in seeking further explanation, which I could not do better than by reading the despatch I had received from your Lordship upon the subject.

       "After hearing the contents of that paper the Prince de Polignac said, that, having in the first instance made it known that the expedition is not undertaken with a view to obtain territorial acquisitions, he had not thought it necessary to insert the positive declaration which your Lordship appears to expect, but that he can have no difficulty in giving me any assurance which may be calculated to remove the uneasiness of His Majesty's Government ; though he begged me to remember that he does not mean to abandon the establishments of La Calle and Bastion de France, together with the rights which have accompanied the possession of those ports during more than a century, and which are necessary for the protection of the French fisheries on that coast. He added that His Most Christian Majesty does not dispute the sovereignty of the Porte, and will not reject the offer of that Government to interfere for the purpose of obtaining the redress he is entitled to expect ; and although the presence of a French negotiator has been rendered impossible by the conduct of the Algerines, if the agents of the Porte can obtain conditions which he has told me the French Government are determined to exact, there will be no reason to send the expedition ; but that they have too much experience of the utter inability of the Porte to influence the authorities at Algiers to induce him to advise his sovereign to delay measures for obtaining this object by force.

" His Excellency further observed, that since France seeks no territorial advantages, in case the present Government of Algiers shall be overturned, the arrangement for the settlement of the future system by which the country is to be ruled will, of course, be concerted with the Sultan, and, being executed under his authority, will imply a due consultation of his rights.

" He has assured me that the instructions for the conduct of the expedition had not yet been drawn up, and that he had hitherto sought in vain for the communications which had been addressed to the French Government upon the departure of Lord Exmouth, in order to follow the precedent established by my Court upon that occasion.

" His Excellency asked me officially to communicate to him your Lordship's despatch, or the substance of that despatch, in a note which he might lay before the king. I did not, however, feel myself at liberty to comply with this request without a special instruction to that effect.

" I have the honour to be, etc.,

(*Signed*)   " STUART DE ROTHESAY."

THE EARL OF ABERDEEN, K.T.
      etc. etc. etc.

In addition to stating to Lord Stuart de Rothesay the precise contrary of the resolution adopted by the French Cabinet on the subject of Algiers, M. de Polignac went so far in his next conversation as to promise equal benefit to the commerce of all nations as an effect of the Algiers expedition.

" Paris, April 9, 1830.

" MY LORD,

" I have to acknowledge your Lordship's despatch No. 18, by which I observe that your Lordship attributes the delay of the French Government to give a full explanation of the objects of the projected expedition, and the assurances that they entertain no views of territorial aggrandizement on the coast of Africa, to my omission to deliver the copy of your Lordship's despatch, No. 13, at the time that document was read to the Prince de Polignac.

" If this excuse has been put forward, I must observe the despatch *in extenso* was read to and read by Monsieur de Polignac when it reached my hands a fortnight ago ; that at least a week has elapsed since the copy was delivered to His Excellency, during which time I have more than once asked him if the explanation required by His Majesty's Government has been rendered the subject of a communication which he promised me on both occasions to send to M. de Laval.

" As His Excellency did not render his compliance with that request dependent on the delivery of the copy of your Lordship's despatch, which, having read, he knew I could have no reason for withholding, and as he told me the explanation required would be sent to Monsieur de Laval in the same form with his earlier communications on this subject, I am justified in rather believing his assertion to be true, that his numerous occupations are the causes of delay.

" In a conversation that took place on the same subject to-day His Excellency said that the explanation had been drawn out in terms which he feels convinced will be considered satisfactory ʰʸ my Government, and that it would have been sent off to-night if he had not thought proper to place it before the Cabinet at their meeting to-morrow morning.

" He hinted that the anxiety which I manifested upon the subject of this explanation indicated a feeling of mistrust which the verbal explanations I had more than once received do not warrant ; that he had looked with satisfaction to this measure, because it offered an opportunity of showing to the world the mutual confidence of the two Governments ; and that he hopes to be consoled for the disappointment he felt in not obtaining the concurrence of my Court in the active operations they are about to undertake, for a purpose of equal benefit to the commerce of all nations, by our concurrence in the future settlement of the questions to which the success of their efforts would give rise; and that, if I had full power, he would readily sign a convention recognizing every principle which had been put forward by my Government in the communications which had taken place on the subject.

"I answered that, feeling no inclination to discredit these assurances, I shall be happy to recognize in the explanatory letter which he has drawn out the proof of his sincere determination to remove every ground for the unfavourable feelings of which, I think unreasonably, the French Government are inclined to complain,

" I have, etc.,

(*Signed*) "STUART DE ROTHESAY."

THE EARL OF ABERDEEN, K.T.
   etc. etc. etc.

Can the French occupation of Algiers be said to have been of equal benefit to the commerce of all nations?

On April 11, 1830, an Order in Council appointed M. de Bourmont, Minister of War, to the command of the expeditionary forces. The Duke of Ragusa, Wellington's adversary in the Peninsular War, was much mortified at finding himself passed over.

On reading the correspondence between London and Paris, seventy years after the despatches were signed, it is hard to see any reason why the inquiries of the British Cabinet should have given offence in Paris. But they did give offence; and Lord Aberdeen, while reasserting his right to make those inquiries, disclaimed all unfriendly intentions in the following despatch:—

" *Foreign Office, April* 21, 1830.

" MY LORD,

  " The French Ambassador has read to me, by order of his Court, a despatch which had been addressed to His Excellency for the purpose of affording to His Majesty's Government those additional explanations respecting the

H

expedition to Algiers which, from the assurances of M. de Polignac, they had been led to expect.

"The Duc de Laval did not feel himself authorized to leave with me, for the information of His Majesty's Government, a copy of this despatch, and he has written to demand the pleasure of his Court before he shall comply with my request.

"I have availed myself of this opportunity to direct the attention of the Ambassador of His Most Christian Majesty to several points adverted to in the despatch in question, as well as to the general tone of remonstrance and complaint in which it appears to be drawn up. It would be difficult to receive a communication of this nature without some reply being thought necessary on the part of His Majesty's Government—a circumstance which, upon such an occasion, it is obvious, had much better, if possible, be avoided ; but as the Duc de Laval has engaged to bring these points under the consideration of his Government, I abstain from entering upon the subject here, and rather confine myself to the statement of some general reflections, which I have to request that your Excellency will submit to M. de Polignac with as little delay as possible.

"The French Government appear to mistake the motives which have induced us to ask for explanations more precise and explicit than those which we have hitherto received respecting the expedition against Algiers. They appear also to have formed an erroneous estimate of the real situation of this country, and to have regarded as evidence of ill-will, of suspicion and distrust, a conduct which has been dictated by a plain sense of duty.

"His Majesty's Government are so far from entertaining these hostile feelings, that they have always been desirous of seeing the most ample reparation exacted from the State of Algiers, and that the efforts of the French Government should succeed in obtaining for His Most Christian Majesty all the satisfaction which His Majesty might justly expect in consequence of the repeated insults and injuries which he had experienced.

"Your Excellency has further been informed that if, in prosecution of this object, His Most Christian Majesty should be enabled to effect the total destruction of piracy, of Christian slavery, and of the imposition of tribute by the Regency of Algiers upon Christian states, it could not but be regarded with satisfaction by the king, our master. These are objects which His Majesty has himself endeavoured to accomplish, and the full attainment of which must be applauded by all Christendom.

" The views to which I have now referred, although sufficiently extensive, are perfectly definite and intelligible. But let us be candid ; the war carried on by France against Algiers is of another character, and a different end is proposed. We are told to prepare for the utter destruction and annihilation of the state itself. Under such circumstances, and with a result of this nature deliberately contemplated, is it unreasonable to expect from the French Government something more than a general assurance of disinterestedness, and an engagement to consult their allies before the future fate of the Regency will be finally decided ?

" A French army, the most numerous, it is believed, which in modern times has ever crossed the sea, is about to undertake the conquest of a territory which, from its geographical position, has always been considered as of the highest importance. No man can look without anxiety at the issue of an enterprise, the ultimate objects of which are still so uncertain and undefined.

" It is to be presumed that His Majesty's Government will not be suspected of culpable indifference to the welfare of the illustrious House of Bourbon. Our desire must be that the result of this expedition may prove to be favourable to their happiness and to the stability of their throne. But if we could so far forget what is due to our sovereign and to ourselves as to rest satisfied with vague explanations in a matter so deeply affecting the interests of British commerce, as well as the political relations of the Mediterranean states, it is certain that the people of this country would not hesitate to pronounce the most unequivocal condemnation of our conduct.

" The views of the French Government in this undertaking being pure and disinterested, it is difficult to conceive that M. de Polignac should experience the slightest reluctance in giving the most satisfactory explanation, or that any false notions of dignity and self-respect should operate to prevent him from doing that which, upon reflection, must appear reasonable.

" In making these observations to M. de Polignac, your Excellency will not only expressly declare yourself to be a stranger to all unfriendly feelings, but you will also disclaim any desire to assume an unbecoming tone, or in any manner to wound the dignity of the French Government.

"Our expectations, we think, are justified by the importance of the occasion, and by the relations of the confidence
existing between the two Governments.

"We ask nothing which, under similar circumstances, we
should not ourselves be ready to grant.

"Your Excellency is authorized to read this despatch to
M. de Polignac.

"I am, etc.,

(*Signed*)   "ABERDEEN."

LORD STUART DE ROTHESAY, G.C.B.
   etc. etc. etc.

There is nothing in this despatch that could give offence to
a reasonable Government. It was crossed by Lord Stuart
de Rothesay's note of April 23, which was followed by a
longer communication, dated the following day. The latter
despatch set forth once more M. de Polignac's positive
disavowal of the aims of the king's Government.

(EXTRACT.)

"*Paris, April* 23, 1830.

"I enclose the copy of a decree, appointing General Bourmont to the command of the expedition against Algiers.

"The communication upon the subject of this undertaking,
which Monsieur de Laval received orders to transmit to His
Majesty's Government, has been very quickly followed by the
publication of a document (an article in the unofficial part
of the *Moniteur* of the 20th April, 1830), to which a very
slight modification of form would give the character of a
manifesto, explaining the causes of the quarrel, and the
intentions of the French Government, in case their arms should
be successful.

"I asked M. de Polignac if this paper is to be considered
official, and especially if it does not claim a larger extent of
territory than has been hitherto understood to be comprised
within the limits of the concession which His Most Christian
Majesty is entitled under treaty to possess.

"He answered, that the article contains no point which is not
recognized by the Government to be perfectly correct ; and
he justified the particular paragraph to which I called his
attention by a reference to engagements antecedent to the

establishment of the present Government at Algiers, which have been repeated and confirmed in subsequent treaties by the Porte and by the local authorities.

(*Signed*) "STUART DE ROTHESAY."

THE EARL OF ABERDEEN, K.T.
    etc. etc. etc.

"*Paris, April* 24, 1830.

" MY LORD,

    " The messenger Latchford arrived last night, and I saw Prince Polignac this morning, when I did not fail to execute your Lordship's instructions.

    " After hearing the contents of your Lordship's despatch of the 21st inst., his Excellency observed that the doubts which have arisen respecting his views in Africa are to be traced to the course His Majesty's Government have pursued ; for that the objection to the measures which had been, in the first instance, contemplated through the agency of Mehemet Ali, had compelled them to make preparations which he is aware must excite the attention, if not the jealousy, of every state interested in the political relations of the countries upon the shores of the Mediterranean.

    " I answered, that since the intervention of Mehemet Ali would have involved other considerations not less embarrassing than the questions which at present occupy our attention, further discussion upon that subject would lead to no satisfactory result ; and that I therefore requested him merely to let me know, whether the reasoning in the despatch I communicated will induce him to give orders to M. de Laval to transmit to your Lordship the assurance, in a written form, which my Government is entitled to expect, that the Court of France entertains no project of conquest or acquisition of territory on the coast of Africa.

    " He said that this assurance was distinctly contained in the despatch which M. de Laval had read to your Lordship, which declares that France will not retain possession of the town or of the Regency of Algiers, though they insist upon the restoration of the establishments they possessed at the period of the rupture ; and that if this declaration has not been conveyed to your Lordship in writing, the omission will be remedied without delay, for that M. de Laval will receive orders to give your Lordship a copy, either of the whole despatch, or of that part of it which your Lordship shall consider most to the purpose.

"I told him that although I justly appreciated the value of his assurances, he must be aware that an abrogation of the projects which he must expect to learn are attributed to the French Government will be more satisfactory if conveyed in a concise form, and not weakened by a reference to questions of indemnity.

"He said that, without complaining of my inquiries, the susceptibility they betrayed was the true excuse of his allusion to a pecuniary indemnity; that otherwise the French Government might in future be exposed to the reproach that such intentions had not been made known; and that I must not therefore be surprised if, in the further communication through M. de Laval to which my representations would give rise, the indemnity would be one of the points which he will be directed to bear in mind.

"The result of my interview, however, enables me to assure your Lordship that the despatch which has already been read to your Lordship will be communicated *in extenso* or in part, as may be deemed most expedient, accompanied by the more precise denial of any view of conquest or of acquisition in Africa than has been hitherto transmitted to His Majesty's Government.

"I have, etc.,

(*Signed*)   "STUART DE ROTHESAY."

THE EARL OF ABERDEEN, K.T.
    etc. etc. etc.

By the last day of April 1830, Lord Stuart de Rothesay had made up his mind that the French Cabinet was not acting straightforwardly with England.

LORD STUART DE ROTHESAY TO THE EARL OF ABERDEEN.

"*Paris, April* 30, 1830.

"MY LORD,

"Since the departure of M. de Bourmont and Monsieur d'Haussez, the business of the Departments of War and Marine has been placed in the hands of the respective Under-Secretaries, under the control of the President of the Council, which duty takes up so much time that he can only communicate with me one day in each week.

"This arrangement did not permit me to see Prince Polignac until yesterday, when I was surprised to learn that he had not sent the orders to M. de Laval to give your Lordship the copy of the despatch in which he proposed to develop the views of the French Government in Africa, and that the further explanations he had likewise promised had not been sent off.

" He said that the king had directed him to write a fuller explanation, containing an account of the several questions at issue with the Regency of Algiers, and more precisely indicating the intentions of the Government, in case the result of the expedition shall be as favourable as the magnitude of the preparations entitles them to expect.

" My answer that a very concise assurance that they entertain no views of conquest and territorial acquisition would save much trouble, and would be more satisfactory to His Majesty's Government, was met by the observation, that our anxiety to prevent the participation of the Pacha of Egypt in the undertaking was too recent a proof of the susceptibility of the British Cabinet upon this question to allow them to hazard the possibility of misinterpretation by the omission of a particle of the explanation which we may be supposed to have a right to expect.

" He then said that our objection to the negotiation with that chieftain had alone induced him to give way to the plans of M. de Bourmont to reduce Algiers by the expedition they had prepared, that he had shown his anxiety to prevent the possibility of the Viceroy's future interference by writing and publishing the enclosed letter to the Chamber of Commerce of Marseilles, and that he is willing to do as much respecting all other points upon which I may manifest uneasiness, though he must hope that our susceptibility will not push him to declarations which the opponents to the present Ministry may be enabled to render grounds of attack upon his Government.

" Though I felt unwilling, after these assurances, to show more than reasonable distrust, I cannot think that the sort of generality by which his expressions are qualified are quite satisfactory. When I objected to conquest and military possession, I observed that his denials were accompanied by the manifestation of the determination to recover the property which he says the French Government have lost at Algiers, and the necessity of preventing future attempts to take it from them. He harped also upon the intention of obtaining the pecuniary indemnity which the country can afford.

" Under such circumstances, as I shall not see His Excellency again until the further explanation he promises shall have been sent away, I think I cannot, in conscience, anticipate that your Lordship will be contented with the result.

" I have, etc.,

(*Signed*) " STUART DE ROTHESAY."

THE EARL OF ABERDEEN, K.T.
     etc. etc. etc.

At this point it may reasonably be inquired whether—all questions of ethics apart—it would not have been simpler and easier for M. de Polignac to have announced his intentions without reserve. It was a plain question between France and England, and between France and England only. It was open to M. de Polignac to argue that the expedition of Lord Exmouth had been only partially successful; insomuch as that expedition had been immediately followed by acts of piracy in the narrow seas. He might have gone on to say that it was undignified for Europe to lie under the obligation of undertaking some such expedition every ten years, which clearly was the only way in which the Barbary States could be kept in order. He might have added that France and England were the only two Powers who were capable of undertaking police duty on so extensive a scale, and that these two Powers once in accord, the other Powers of Europe, who could not protect themselves, could have no legitimate grounds of complaint at their decision. One is bound to admit, however, that such a bargain could not have been in accordance with the traditions of British diplomacy, and England must, perhaps, rest contented to bear some part of the blame for M. de Polignac's insincere attitude.

On May 4, 1830, Lord Aberdeen wrote a strong despatch to Lord Stuart de Rothesay :—

"*Foreign Office, May* 4, 1830.

" MY LORD,

    " The delay which has taken place in furnishing your Excellency with more precise and official explanations respecting the ulterior projects of the French Government in their expedition against Algiers has been observed with much concern. The assurance of M. de Polignac that these explanations would be promptly afforded have been so positive, and so frequently repeated, that His Majesty's Government are at a loss to comprehend by what justifiable motives the delay has been produced. The affair, in truth, begins to wear a sinister appearance, and to give rise to doubts and suspicions which it would be very far from the desire of His Majesty's Government to entertain.

"M. de Polignac expresses a hope that our expectation may not be so unreasonable as to force him to declarations which must prove injurious to the Government of His Most Christian Majesty. It can scarcely be necessary for your Excellency to assure the French Minister that such a result could not be contemplated by us with any degree of satisfaction. The whole character and language of my despatch of 21st April, which you were directed to read to M. de Polignac, sufficiently attest the cordial and friendly feelings of His Majesty's Government. But we have a duty to perform from which we cannot shrink. It is clearly our duty to require an official explanation of the designs of the French Government in equipping and fitting out a military expedition of unexampled magnitude, and thereby calculated to excite speculation and apprehension throughout the south of Europe. From our confidential relations with the Court of France we are entitled to receive this information, which is so much the more due in consequence of the conduct observed by the British Government on a similar occasion. Your Excellency cannot be ignorant that the language of persons possessing much influence in France, and of those nearly connected with the Government, is very much at variance with the verbal assurances which you have received, and therefore renders some official explanation more indispensable. If the projects of the French Cabinet be as pure and disinterested as is asserted by M. de Polignac, he can have no real difficulty in giving us the most entire satisfaction. A concise and simple declaration would not only answer the purpose better, but it would appear to be more natural than the course which your Excellency states that the French Minister has been commanded by His Most Christian Majesty to adopt. To envelop in much reasoning, and to mingle considerations of national dignity and punctilio with the statement of intentions such as I have mentioned, appear less calculated to produce conviction and to convey the impression of sincerity and frankness.

"Should the promised explanations not yet have been forwarded to the French Ambassador at this Court, your Excellency will endeavour to see M. de Polignac without loss of time, and you will represent to him the serious effects of further delay. After all that has passed, the French Minister cannot be surprised if injurious suspicions should be created and confirmed ; and he must be aware that he will make himself responsible for the consequences, however unfortunate, which may attend a state of distrust and apprehension.

"I am, etc.,

(*Signed*) "ABERDEEN."

LORD STUART DE ROTHESAY, G.C.B.
    etc. etc. etc.

Here is language, if not of menace, at least of profound mistrust. "The affair, in truth, begins to wear a sinister appearance ; " " serious effects of further delay " in furnishing "promised explanations"; " consequences however unfortunate." This is alarming language. What lies behind ? It may, perhaps, be noted that M. de Polignac had thought fit to communicate his despatch of January 10, 1830, in confidence to St. Petersburg before forwarding it to the other Great Powers. It may be well also to note that after the revolution of July, 1830, the attitude of Russia towards England and France changed materially. The British Cabinet may be presumed to have been informed in detail of the temper that lay behind this change of attitude. The Cabinet could have been in a position to judge whether the understanding between the Russian Government and the Ministry of Charles the Tenth was inconsistent with the interests of Great Britain. It is certain that this understanding ceased from the date of the accession of Louis Philippe ; that, coincidently, the angry correspondence between London and Paris was entirely dropped, and that a policy of silent acquiescence in the French occupation of Algiers was adopted by England. This much is certain ; beyond is speculation.

In later years (August 7, 1844) Lord Palmerston declared in the House of Commons that in 1830 the Cabinet had acquiesced in the French occupation of Algiers, in order that the Ministry of Prince Polignac might be maintained in power. But in so far as the attitude of the British Cabinet was influenced by the personality of the French Prime Minister, it appears that it was precisely the presence of Prince Polignac in the French Cabinet that aroused suspicion. Whether it was that his personality was disagreeable, or that he represented a policy that was positively alarming, the fact remains that from the date of his downfall the correspondence between the two Cabinets lost all its bitterness. Presumably his policy disappeared with him. Presumably, also, that policy included designs beyond the mere annexation of Algiers—designs which the British Cabinet more than suspected, and which they were almost prepared to resist by force of arms.

To return to the year 1830. On April 19th, M. de Bourmont had left Paris to take up his command. As regarded the point at issue between France and England there was nothing more to be said after Lord Aberdeen's despatch of May 4. Since April 15, however, a new complication had arisen ; bulletins relating to the king's health were published from this date, and by the middle of May it became known that a demise of the Crown might shortly be anticipated. The menace in the last sentence of the following despatch of May 11 was therefore probably discounted by the French Foreign Office.

"Foreign Office, May 11, 1830.

" By the despatch of 23rd March your Excellency was informed of the reason which had induced His Majesty's Government to seek for some more precise and official explanation of the ulterior objects of the expedition, in addition to that which had been clearly communicated in this form by the Duc de Laval.

" In your despatch of April 9th your Excellency observes, that M. de Polignac had assured you that the explanation required would be sent to M. de Laval in the same form as the preceding ; and in the course of the conversation the French Minister strikingly illustrated his desire to satisfy His Majesty's Government by declaring, that if you had full powers, he would readily sign a convention recognizing every principle which had been put forward by your Government in the communications which had taken place on the subject.

"In your despatch of April 24th, in answer to a question from your Excellency, whether M. de Laval would be authorized to give the additional assurances in a written form, you describe M. de Polignac to state that the French Ambassador could be instructed to communicate to His Majesty's Government, either the whole despatch from his Court, or such part of it as should be considered by His Majesty's Government most to the purpose.

"In your despatch of April 30th, throughout the whole narrative of your conference with M. de Polignac, it is clearly implied that this communication was to be made in a written form ; and the observations of the French Minister show an anxiety to prepare the statement in such a manner as to give satisfaction to the British Government.

"The request, indeed, contained in my despatch of March 23rd of an official assurance, necessarily precludes any other mode of communication ; and as you were instructed to deliver a copy of that despatch to M. de Polignac, it is not possible that the French Government should have laboured under any misconception.

"Your Excellency will not fail to draw the serious attention of the President of the Council to the promises which you have received, and the pledges repeatedly given, as well as to the manner in which it is now proposed that they should be redeemed.

"When you shall have reported the result of the appeal thus made to the consistency and good faith of M. de Polignac it will be my duty humbly to take His Majesty's commands respecting such further instructions to your Excellency as the occasion may require.

(Signed) "ABERDEEN."

LORD STUART DE ROTHESAY, G.C.B.
etc. etc. etc.

The next day the Prince de Polignac addressed to the Duc de Laval the following despatch—

"Paris, May 12, 1830.

"MONSIEUR LE DUC,
"At the moment when the fleet which conveys our army to Africa is leaving France, the king feels the necessity

of making known to his allies how sensible he has been of the marks of interest and friendship which he has received from them during the important conjuncture of circumstances which preceded the departure of the expedition against Algiers. His Majesty has applied for their concurrence with perfect confidence ; he has treated, it may be said, publicly a question which he has thought fit to make common to all Europe ; his allies have responded to his confidence, and they have afforded him sanction and encouragement, the remembrance of which will never be effaced from his mind.

" To make a return for conduct so loyal and friendly, His Majesty is now desirous of laying before them again, at the moment of the departure of the French fleet, the object and aim of the expedition which he is sending against the Regency of Algiers.

" Two interests, which by their nature are distinct, but which are closely connected in the mind of the king, have led to the armaments which have been prepared in our ports. The one more especially concerns France ; it is to vindicate the honour of our flag, to obtain redress of the wrongs which have been the immediate cause of our hostilities, to preserve our possessions from the aggressions and acts of violence to which they have been so often subjected, and to obtain for us a pecuniary indemnity which may relieve us, so far as the State of Algiers will allow, from the expense of a war which we have not provoked ; the other, which regards Christendom in general, embraces the abolition of slavery, of piracy, and of the tribute which Europe still pays to the Regency of Algiers.

" The king is finally resolved not to lay down his arms, or to recall his troops from Algiers, until this double object shall have been obtained and sufficiently secured ; and it is with the view of coming to an understanding as to the means of arriving at this end, so far as regards the general interests of Europe, that His Majesty on the 12th of March last announced to his allies his desire to take measures in concert with them, in the event of the dissolution of the Government actually existing at Algiers, in the struggle which is about to take place. It would be the object of this concert to discuss the new order of things which it might be expedient to establish in that country for the greater benefit of Christendom. His Majesty thinks it right at once to assure his allies that he would enter into those deliberations prepared to afford all the explanations which they might still desire— disposed to take into consideration the rights and interests of all parties, himself unfettered by any previous engagements—at liberty to accept any proposition which might be considered proper for the attainment of the object in question, and free from any feeling of personal interest; and as the state of things foreseen by His Majesty may very shortly be realized, if Providence deigns to protect our arms, the king now invites his allies to furnish their Ambassadors at Paris with contingent instructions on this subject.

"You will have the goodness, M. le Duc, to make this proposition to Lord Aberdeen ; and if that Minister wishes it, you will give him a copy of this despatch.

"Accept, etc.,

(*Signed*) "LE PRINCE DE POLIGNAC."

M. LE PRINCE DUC DE MONTMORENCY.
        etc. etc. etc.

The circular note herein mentioned, as of date March 12, is here inserted for reference.

PRINCE POLIGNAC TO THE DUC DE LAVAL.

"*Paris, March* 12, 1830.

"MONSIEUR LE DUC,

        "When we communicated to our allies the destination of the armaments now preparing in the forts of France, we spoke of the results to which they might lead, with a reserve which appeared to us to be called for by the uncertainty of the chances of war. Many Cabinets having since invited us to declare to them, in a more precise manner, the object which we propose to attain by our expedition against the Regency of Algiers, His Majesty is pleased to comply with this desire, so far as depends upon him ; and he authorizes me to give to the several Cabinets the following explanations ; you may address them, M. le Duc, to the Government of His Britannic Majesty.

"The public insult offered by the Dey to our Consul was the immediate cause of a rupture, which was moreover but too well justified by numerous infractions of treaties, by the violation of rights which a possession of many ages' duration had consecrated, and by the injury done to interests of very high value and importance.

"To obtain satisfaction for the insults offered to one of his agents, suitable reparation for the injuries experienced by

France, and the performance of the engagements which the Dey refused to fulfil—such was at first the object which the king proposed to attain.

"Events have subsequently given a more extended development to the projects of His Majesty.

"The Dey has ruined and utterly destroyed all our establishments on the coast of Africa; a three years' blockade has only increased his insolence, and instead of the reparation due to us, he has spoken only of claims and pretensions which he himself reckoned upon making good against France. In short, he has replied to the pacific propositions, which one of the commanders of our navy was sent to convey to him, even in his own palace, by an absolute refusal; and at the moment when the vessel employed for the negotiation, and carrying a flag of truce, was preparing to leave the port, it was suddenly attacked by the fire of all the nearest batteries, upon a signal given from the very castle which was occupied by the Chief of the Regency.

"The king, M. le Duc, has therefore been compelled to acknowledge that no arrangement could be practicable with the Dey, and that even if it should be possible to induce him to conclude any treaty whatsoever, the previous conduct of the Regency, compared with more recent events, left no security that such an arrangement would be better observed than our conventions, so often renewed and so often violated by the Algerine Government.

"These considerations have convinced us of the necessity of giving a more extended development to the war. From that period also it became incumbent upon us to consider how to give to this war an object, the importance of which would correspond with the extent of the sacrifice which it would impose upon us; and the king, no longer confining his projects to obtaining reparation for the particular wrongs of France, determined to turn to the advantage of all Christendom the expedition for which he was ordering the preparations to be made, and His Majesty adopted as the object and recompense of his efforts—

"The complete destruction of piracy.

"The total abolition of Christian slavery.

"The suppression of the tribute which Christian Powers pay to the Regency.

"Such, if Providence assist the arms of the king, will be the result of the enterprise for which preparations are now making in the ports of France. His Majesty is determined to prosecute it by the employment of all the means which may be necessary to secure its success ; and if, in the struggle which is about to take place, it should happen that the existing Government at Algiers should ever be dissolved, in that case, Monsieur le Duc, the king, whose views upon this important question are perfectly disinterested, will concert with his allies for the purpose of deciding what shall be the new order of things which may be substituted, with the greatest benefit to Christendom, for the system which has been destroyed, and which may be best calculated to secure the triple object which His Majesty proposes to attain.

"You may convey these communications, M. le Duc, to the knowledge of the Government of His Britannic Majesty ; and if Lord Aberdeen wishes to have a copy of the present despatch, the king authorizes you to give it to him.

"Accept, etc.,

(*Signed*) "LE PRINCE DE POLIGNAC."

On this Lord Stuart de Rothesay wrote—

"*Paris,* May 14, 1830.

"The Prince de Polignac yesterday read to me a despatch, which he has addressed to the several Ministers accredited to the Courts in alliance with France, containing a further exposure of the objects of the expedition.

"The copy of this despatch will, he tells me, be delivered to the Ministers by whom it may be required.

"It was certainly the desire of M. de Polignac to confine the operations of this country against Algiers to the employment of their navy, leaving the attack by land to be wholly executed by the Viceroy of Egypt, and he was induced to abandon this project by the remonstrance of His Majesty's Government, joined to the arguments of his colleague for the War Department.

(*Signed*) "STUART DE ROTHESAY."

THE EARL OF ABERDEEN, K.T.
etc. etc. etc.

And again on the same date at greater length—

"*Paris,* May 14, 1830.

"I have received your Lordship's despatch of 12th May, pointing out the manifest contradiction between the explanations on the subject of the French expedition to Algiers, contained in a letter which has been read to your Lordship by the Duc de Laval, and the assurances of the French Minister to myself, if I have not misrepresented the language of M. de Polignac in my despatches.

"I could not more distinctly execute your Lordship's instructions in asking an explanation of this contradiction, than by reading the contents of this despatch to Prince Polignac, reminding His Excellency upon what occasions he had authorized me to convey to my Government the assurances to which your Lordship adverts, and repeating to His Excellency the observation, which I had brought forward in several conversations, that these assurances are inconsistent with the course to be pursued.

"His Excellency did not deny that the considerations which had arisen between the period when he had conferred with me, and that when he sent off these despatches to M. de Laval, has led to some variation between the form as well as the nature of his verbal and written explanation, but that the time was fast approaching when I should be compelled to admit the truth of all the assurances I had received ; that if upon the appearance of the expedition before Algiers, the Dey shall consent to the terms proposed, their immediate return to France will put an end to every question ; while if the resistance of the Algerines shall lead to a struggle which terminates in the dissolution of the Government, that the measures to be adopted for the resettlement of the country, whether by placing it under the rule of a Turkish Pacha, or such other arrangement as may be thought expedient, will be concerted in a conference of the representatives of the allies, and not exclusively decided by the French Ministers ; and that the general commanding the expedition had therefore received orders not to commit his Government by any engagement which can stand in the way of the resolution.

"After this statement he cannot conceive that more explanation than has been contained in the despatch which M. de Laval was yesterday directed to deliver to your Lordship will be required.

(*Signed*)   "STUART DE ROTHESAY."

THE EARL OF ABERDEEN, K.T.
    etc. etc. etc.

By May 24, the king could no longer sign his name and

I

a stamp was permitted by special Act of Parliament, to be used for the transacting of important public business.

A week later Lord Aberdeen replied to the circular of May 12.

"*Foreign Office, May* 31, 1830.

"MY LORD,

"I enclose to your Excellency the draft of a Note which your Excellency will address to the Prince de Polignac, in answer to the official communications which have been made by the Ambassador of His Most Christian Majesty at this Court to His Majesty's Government respecting the French expedition against Algiers.

"I am, etc.,

(*Signed*)   "ABERDEEN."

LORD STUART DE ROTHESAY, G.C.B.
etc. etc. etc.

Note presented to the Prince de Polignac by Lord Stuart de Rothesay, dated June 3, 1830.

"The undersigned has received instructions to lay before the Cabinet of the Tuileries the following observations in answer to the official communications which have been made to his Court respecting the expedition of a French force against Algiers.

"The cabinet of the Tuileries is no stranger to the sentiments which have been constantly entertained and frequently expressed by the British Government upon this subject. The undersigned is now commanded to repeat that the king his master has long been sensible of the injuries sustained by His Most Christian Majesty from the Regency of Algiers, and he has always expected that such injuries would be duly avenged.

"If, in exacting reparation for outrages committed against himself, His Most Christian Majesty should be enabled entirely to put an end to the evils of piracy and of Christian slavery, the benefit must be acknowledged by all Christendom.

"In case it should be found impracticable to attain these objects without the total subversion of the Algerine State, His Most Christian Majesty has desired to receive the opinion and counsel of his allies respecting the manner in which this conquest might be rendered most advantageous to the general interests of Europe. The repeated disavowal of all projects of ambition and aggrandizement made by the Prince de Polignac, and the assurances which have been received from the Ambassadors of His Most Christian Majesty in London, forbid the suspicion of any design on the part of the French Government to establish a permanent military occupation of the Regency, or to accomplish such a change in the state of territorial possession on the shores of the Mediterranean as should affect the interests of European Powers.

" The undersigned cannot avoid calling the attention of the Prince de Polignac to the peculiar situation of Algiers in its relation to the Ottoman Porte. Various governments of Europe have contracted engagements with the Regency, as with an independent state, and in virtue of conventions with the Porte to that effect, have made the Algerine rulers responsible for the acts of their subjects. Other Powers continue to regard the Barbary States as essentially dependent on the Turkish Empire, and claim accordingly from the Turkish Government compensation and indemnity for all injuries received from these States. The supremacy of the Sultan is admitted however by all ; and His Most Christian Majesty himself has only recently renounced the hope of reconciling his differences with the Regency by means of the intervention of the Porte. A Turkish Commissioner has actually arrived at Toulon, having been prevented by the French blockading squadron from landing at Algiers, whither he had been sent from Constantinople in order to enforce compliance with the just demands of the French Government.

" If the main object of this expedition should be the conquest of Algiers, rather than the reparation of injuries, and the chastisement of the Regency, the undersigned would submit to the serious consideration of the Prince de Polignac, what must be the effect of a precedent which thus disposes of the rights of a third party against whom no complaint whatever has been alleged."

In the meantime, the suzerain of the Dey of Algiers had at last, after years passed in idle contemplation of the dealings of European Powers with his vassal, given signs of life. The step which was now taken by Sultan Mahmúd, if taken earlier, even six months earlier, might have had important results for the Ottoman Empire. The Sublime Porte determined to mediate between France and Algiers.

This excellent resolve took shape in the mission of an important Turkish official, Tahir Pacha, holding the rank of admiral. Tahir Pacha was despatched to Algiers, charged with instructions of the most conciliatory nature : but, since no agreement as to his reception had been arrived at between France and Turkey, the commandant of the blockading squadron declined to allow him access to the city. This was a disheartening reception for the peace-maker, but he was escorted towards Toulon by a vessel of the blockading squadron. This had occurred on May 21, 1830. Five days later the peace-maker and his escort encountered the entire Mediterranean fleet of France, escorting the Count de Bourmont and the army of conquest. Tahir Pacha was too late : he could do nothing but remonstrate. Probably the result would not have been different had he arrived earlier.

" Paris, May 31, 1830.

"So soon as the telegraphic despatch announcing the arrival of Tahir Pacha at Toulon came to my knowledge, I lost no time in asking the Prince de Polignac to explain to me the cause of that event.

"His Excellency said that the telegraphic despatch contained all the information he had received, by which it appeared that the Turkish vessel in which Tahir Pacha had taken his passage to Algiers, having been repulsed from that port by the blockading squadron, had steered for the French coast, and had met with the expedition the day after they sailed. Tahir Pacha had gone on board the Admiral's ship, and after a long conference with Count Bourmont, he determined to continue his course to Toulon, where he announced that he was the bearer of propositions to the French Government, and that a letter to that effect had been immediately sent off, but had not been hitherto delivered.

"Prince Polignac declared his utter ignorance of the tenor of this communication, though he does not seem to doubt that Tahir Pacha will think it expedient to proceed to Paris.

"I observed to His Excellency, that, however imperfectly I had learned the object of the voyage of this personage, I knew he was directed by the Sultan to use his best endeavours to prevent hostilities, by directing the Dey to submit to every just demand which the French Government is entitled to bring forward; that it appears strange he should not have been allowed to pass the blockading squadron ; and yet more strange that, after communicating with the commander of the expedition, he should not have accompanied them to Algiers, for the purpose of carrying the order of the Sultan to the knowledge of the local authorities ; that at Toulon, he will, without doubt, be detained in quarantine ; and if he intends coming to Paris, he may possibly not reach Algiers till long after it shall be too late to take a part in the negotiations which are likely to follow the capture of the place.

(*Signed*) "STUART DE ROTHESAY."

The expeditionary force destined for the conquest of Algiers numbered 40,000 men. This was too large an army for a punitive expedition ; too small for an army of occupation, as events were to show only too soon. But it was a gallant army, well provided, and reasonably well led ; and it set sail from Toulon on May 25, 1830. Seven little steamboats accompanied the hundred sail that composed the convoying fleet. Vice-admiral Duperré was in command. On June 14 the disembarkation commenced. Two smart skirmishes, one at Staoueli, the other at Sidi-Khalef, hardly hindered the eastward march of the army, which proceeded on June 30 to the attack of the famous " Emperor's Castle."

This was the fortification that marked the site of Charles V.'s encampment on October 23, 1541. It was re-christened Fort Napoleon by the French troops, for whom there was but one emperor. Officially it was known as Sultankalassi, or as Burj Muley Hassan (the tower of Muley Hassan). It was shelled on the morning of July 4, and so effectually shelled that in the course of the afternoon the Dey offered terms. He offered to accept all the conditions which he had previously rejected—reparation to Mr. Consul Deval, etc.—and to pay the expenses of the expedition. This offer was rejected as *dérisoire*, and was afterwards so described. But it is not *dérisoire*: it is a very good first offer, for the first offer in a negotiation of such importance. If the assurances which MM. de Polignac and de Laval had proffered to Lord

Aberdeen had substantially represented the policy of the French Government, such an offer would have paved the way to a pacific settlement. Naturally, it would not have been immediately accepted. Security would have been taken for the payment of the army's expenses. M. de Bourmont would further have stipulated that it was necessary for the honour of the French that the victorious troops should march through Algiers, and that the French flag should be hoisted for six hours on the fortifications of the city. There would have been some attempt to resist these—or similar—additional conditions, but after resistance for form's sake the attempt would have been abandoned, and the expedition would have returned to Toulon, its work over. It is to be observed that (with the addition of a trifling extension of the rayon round Ceuta, and the establishment of a missionary college) these very terms were all that France and England permitted Spain to impose upon Morocco after Marshal O'Donnell's campaign of 1859–60. But, of course, there had not been from the first the slightest intention of confining the activity of the French army to punitive operations.

Lord Stuart de Rothesay, however, seems to have thought it worth while to remind M. de Polignac that the "avowed object" of the expedition had now been obtained.

"Paris, July 16, 1830.

"I saw Monsieur de Polignac within a few hours of the departure of the last messenger.

"I told His Excellency that, so soon as I heard of the complete success of the expedition against Algiers, and attainment of the avowed object of the undertaking, I came to offer him my congratulations, in the conviction that they will keep their faith with my Court; and that, notwithstanding all that has been written and said to the contrary, they will not take advantage of the moment of success to fall from the assurances he has given me, in the name of his sovereign, that the expedition was undertaken for the sole purpose of vindicating the national honour, and not with the view of acquisition or conquest.

"His Excellency answered me by declaring his readiness

to repeat his former assurance, from which he declared that their late success gives the French Government no inclination to depart.

(*Signed*)   "STUART DE ROTHESAY."

The last Dey of Algiers embarked on the *Jeanne D'Arc* on July 10, 1830.   After ten days' quarantine endured at Mahon, he landed at Naples, his destined place of exile, on the 31st, and heard that his conqueror Charles X., the last King of France, was himself an exile.   The "Revolution of July" had broken out on the 19th, only ten days after the cannon of the Tuileries had announced to an indifferent populace that France was mistress of Algiers.   A month earlier George IV. had passed away.   He received the last sacraments from the Bishop of Chichester, but lingered unexpectedly for two or three days, retaining consciousness but not the power of speech.   He died suddenly at three o'clock in the morning of June 26, 1830.

PART II

THE STRUGGLE FOR EGYPT

II

In the history of France and England in the Mediterranean it is natural to pause at the year 1830. It is this year, and not the year 1815, as we might have supposed, that is the year of crisis. Between 1815 and 1830 there are "alarms and excursions," but no new policy is inaugurated. It appears that the nineteenth century is about to carry on the political traditions of the eighteenth. We prepare for, and in fact find, punitive expeditions against the coasts of Africa, congresses and conventions as regards other Mediterranean questions, a state of things in which the Powers are contented to live as formerly from hand to mouth. M. de Châteaubriand, indeed, claims that the Government of the restored royal family of France has achieved three great deeds, viz. the restoration of the Spanish monarchy, the abolition of Christian slavery, and the liberation of Greece. The citations are chiefly remarkable as illustrating the extent to which the political horizon has widened since the days of M. de Châteaubriand. Spanish politics are to-day but eddies in a backwater of history ; and the Spanish campaign, if of service to a dynasty, was hardly of service to the cause of humanity in so far as it strengthened the throne of Ferdinand of Spain. The liberation of Greece and the abolition of Christian slavery in Algiers were undoubted services to the cause of humanity ; but Greece has disappointed her champions, and France cannot claim all the credit for the downfall of Algiers, still less for the battle of Navarino. What is really remarkable about the action of the French Government on this occasion, is the bold inauguration of a policy of annexation in respect of the territory of Islam on the northern coasts of Africa.

It is, then, from 1830 that we must date the new policy of France, and as a result a marked change in the relations of France and England. In the severe diplomatic contest that preceded the surrender of England on this occasion, there are many points that deserve something more than passing attention. The first is that the struggle was, from first to last, between two Cabinets; the people on either side remaining indifferent to the result, and almost ignorant that the struggle was proceeding. The French Cabinet won ; nevertheless their success counted for nothing in the events that culminated rapidly in the immediate downfall of the Ministry and the expulsion of the dynasty. This is perhaps not surprising in France. We have seen something like this in our own days, when smaller triumphs (but still triumphs) like Siam, Tonkin, Tunis and Madagascar have altogether failed to give stability to a tottering Ministry. "The Government" in France has been so long accustomed to regard itself as an entity apart from the nation, that the historic divorce of sentiment between people and Cabinet is hardly remarkable, except on an occasion like the conquest of Algiers—an event of the kind to appeal to Frenchmen, an event striking, dramatic and with a promise of immediate material advantage to French investors.

The French Cabinet, then, won, and gained nothing by its victory. The English Cabinet was defeated, but lost nothing by the humiliation. But if the course of events in France at this epoch is in line with the position taken up in these pages as regards the rivalry of France and England in the Mediterranean, the course of events in England is a direct, a flagrant contradiction of that position. The Cabinet pushes on, the people hangs back ; the Cabinet is defeated and the people, far from censuring the Cabinet, remains indifferent to the results of governmental action. It is submitted that, upon examination, this will prove to be an exception that confirms rather than weakens the general position. The traditional anxiety of England on the subject of the Mediterranean was as nothing in 1830, by the side of the angry demand for domestic reform. England was indifferent to foreign policy because she was intent on social revolution.

The same attitude of mind that marked the policy of England at the commencement of the forty good years of France in the Mediterranean was being constantly strengthened throughout the entire period. The first Reform Bill was hardly passed, when the good work (as the phrase went at the time) was carried on by the Repeal of the Corn Laws. Hardly was the policy of Free Trade inaugurated when the agitation for a further extension of the Franchise was set on foot. These measures took up the attention of the country to the exclusion of questions of foreign policy. The assertive attitude of Lord Palmerston was a personal matter, and existed side by side with an indifference to Imperial matters which was the natural result of ignorance. It would be hard to find another example in history of an empire that for forty years was neglected, and invited to go to pieces ; was even mocked at by its rulers for the extraordinary slowness of mind that hindered its disruption, and caused its constituent parts to cling to the exploded ideas of unity and patriotism, and that yet nevertheless did not break up. During these forty years France made good use of her time in the Mediterranean. Her influence in Egypt was increased until it was almost as effective as a declared Protectorate ; the province of Algiers was gradually subjugated. The history of France in Algiers belongs wholly to the nineteenth century ; but the dealings of France with Egypt began in the eighteenth century, and must be examined at greater length than has hitherto been possible.

We read, with something more than surprise, that Bonaparte designed to reach Mysore through Egypt. Lord Elgin, the British Ambassador at Constantinople, thought it of the first importance to keep Lord Mornington, the Governor-General of India, regularly informed by special Tartar of every change of affairs in Egypt throughout the closing years of the eighteenth century. We marvel that men so favourably placed for acquiring accurate information as Bonaparte and Lord

Elgin should, the one seriously contemplate moving thirty thousand men down the Red Sea and across the Indian Ocean in Arab dhows, and the other be seriously alarmed at the menace. But the French plan for invading India appears comparatively sober, when we learn how wild and how scanty was the available information about a country no more remote than Egypt. Tales worthy of Herodotus, and ignorance such as now exists of no country in the world except, perhaps, Tibet, prepare us for anything. A British cavalry officer gravely records that the prosperity of the province of Dongola is due to the services of a useful animal akin to the mule, a cross between the hippopotamus and the mare. So late as the year 1841 an elementary fact like the distance between Suez and Cairo was imperfectly ascertained. At the date of the Treaty of Amiens Grand Cairo was credited with two millions of inhabitants ; its actual population at that time being two hundred and fifty thousand. Small wonder that Bonaparte, and still less that the Directory, should see nothing miraculous, or even very difficult, in reaching India by way of Egypt, cutting the Suez Canal on the way.

The first step towards this end taken by Bonaparte was the reservation of the Ionian Islands to France by the Treaty of Campo Formio signed October 17, 1797. The details of the treaty provision were to be settled by the conference at Rastadt, from which Bonaparte turned aside. He hurried to Paris, with his mind bent on the conquest of Egypt, and reported himself to the Directory on December 5, 1797. The army of Egypt (still up to the last moment called " the army of England ") set sail from Toulon on May 19, 1798. The open secret between Bonaparte, Brueys, Talleyrand and the Directory was thus kept for six months. The first measure to this end was the appointment of Bonaparte to the command of the army of England, and his despatch to the coasts of Normandy on a tour of inspection of that army. The invasion of England was officially declared to be postponed until the autumn should bring fogs ; and in the meantime the choice between England and Egypt was secretly and daily discussed

by the Directory.   The ambitions of Bonaparte had this draw-
back for the Directory : that, if realized, they would make a too
powerful citizen still more powerful.   But they also had this
recommendation, that they would lead him far away into
unknown perils, and might, whether realized or not, relieve
France altogether of a personality that already threatened to
become overwhelming.   It was finally decided to entrust him
with the command, and his instructions were issued on March
5, 1798.   In the meantime he was ordered to Brest on a
second tour of inspection ; and to keep up the delusion he
wrote to Rastadt announcing his impending arrival there, and
openly prepared to take his place at the conference.   The
following is the text of his instructions :—

"Vous trouverez ci-joint, Général, les expéditions des
arrêtés pris par le Directoire exécutif pour remplir prompte-
ment le grand objet de l'armement de la Méditerranée ; vous
êtes chargé en chef de leur exécution.   Vous voudrez bien
prendre les moyens les plus prompts et les plus sûrs.   Les
ministres de la Guerre, de la Marine et des Finances sont
prévenus de se conformer aux instructions que vous leur
transmettrez sur ce point important dont votre patriotisme a
le secret, et dont le Directoire ne pouvait mieux confier
l'exécution qu' à votre génie et à votre amour pour la vraie
gloire.

(*Sign.*) LA REVEILLÉRE-LEPEAUX, MERLIN, P. BARRAS."

Nothing could be vaguer than this.   The deception was
complete—England imagining that a descent on the coast of
Ireland was projected, and public opinion in France dwelling
on four conjectures—namely, India, Brazil, the Crimea and
Sardinia.

"The left wing of the army of England," as the Egyptian
expedition was called, was composed of the best troops of
France—the army of Italy.   It numbered thirty-six thousand
veterans, and was convoyed in four hundred transports,
guarded by a fleet of one hundred sail, manned by ten
thousand sailors.   It was accompanied by a little band of
scientific men, who were to form the Egyptian Institution,

and whose presence gave a kind of guarantee that something more than a raid was intended. In fact the Directory had been brought to consent to the undertaking by Bonaparte's sanguine estimate of the chances of colonizing Egypt successfully; and it was as colonists rather than conquerors that he addressed the famous army when he took command at Toulon on May 9, 1798.

The Consul-General of the French Republic at Aleppo appears to have been the first person to make the obvious comment that Turkey could not be safely neglected. Before the fatal First of August, and while the expedition was supposed to be in the full tide of success, he wrote anxiously to the Minister of Foreign Affairs. He ought, he said, to have been, at the least, officially authorized to employ conciliatory language when once the expedition was under way. So many disasters befell the expedition that this omission merely contributed to the fatal result. It may have been designed, or it may have been the outcome of ignorance, that ignorance which left the impression that Alexandria was itself the capital of Turkey, and favourable to the French invasion.

In great and natural ill-humour at the overcrowding of the ships, the army made its way eastward. Some anxiety at the possible neighbourhood of Nelson was felt at the start, and still more at the landing, where it was feared that the disembarking troops might be taken at a disadvantage. After the fall of Malta, however, the troops gained confidence, and when the danger was, as he hoped, past, Brueys himself did not hesitate to say that Nelson had avoided a battle. The capture of Malta was held to be of the first importance. The Ionian Islands were very well so far as they went, but they had only been acquired at the price of extensive concessions to Austria. These concessions had made Austria a naval Power; or, rather, they had awakened in Austria the ambition of becoming a naval Power. If, in pursuance of this ambition, Austria should contrive to gain possession of Malta, the French would have been outwitted,

or their occupation of Corfu would be nullified. This was
the view unanimously adopted by those in power in France,
and the first secretary of the Republican Legation at Genoa,
Poussielgue, was entrusted with the duty of inspecting and
reporting, or (as the English bluntly put it) bribing the
Knights of Malta into a favourable attitude of mind.
Hompesch was Grand Master. Poussielgue's duties, or as
the English called them, "intrigues," had commenced as
early as December 1797. Brueys had anchored in Malta
and carefully sounded the passage and anchorage at his
leisure in March 1798. Repeated warnings had reached
Hompesch from Rastadt that the French Republic had
designs on Malta ; notwithstanding which, he refused to
squander the treasures of the Order in any such sacrilegious
work as putting the fortifications into repair. It matters little
whether Hompesch was in earnest or not ; whether he was
merely incompetent, or had really retained, to the verge of
the nineteenth century, a piety that has rather the ring of
the fifth century ; or whether he was already assured of a
principality in Germany and the revenue of £12,000 a year,
which was the price of his complaisance. In two days the
impregnable fortress had ceased to be an independent state,
Caffarelli summing up the situation in the famous phrase,
"It was lucky that some one was inside to open the gates."
Regnault de St. Jean d'Angely was placed in command, and
the actual destination of the army was at last revealed. Off
Alexandria, and from on board the flagship *L'Orient*, the
comment was added, "Vous porterez à l'Angleterre le coup
le plus sûr et le plus sensible en attendant que vous puissiez
lui donner le coup mortel."

On the whole the voyage had been prosperous. It had
lasted a long time, from May 19 till June 30. The fleet
had not been attacked, and the Admiral was genuinely of
opinion that the English were afraid of him. There had been
little or no sickness on board; everything promised well.
The sea and land forces had got on badly with each other.
There had been no actual quarrel, and as the two branches of

the service would not have much work to do in common, their disagreement was not regarded as material. Nevertheless it existed; but was perhaps the only cloud on the horizon when the invading army of Egypt disembarked at Alexandria.

At that time the English interests in Egypt were insignificant. It is true that the post of Consul-General existed, but so far from being the incalculably important post that it has grown to be in the course of the last century, there were doubts in the year 1795, whether it was worth keeping up, in view of the trifling interests that its incumbent was supposed to superintend. The Consul-General was a Mr. George Baldwin, to whom the post had been granted as a provision for his declining years. He was dismissed in March 1796, at which date his pay was four years in arrear. His destined successor was an officer of the East India Company, should they find it worth their while to maintain an agent in Egypt. But if the Egyptian Government paid little attention to a British governmental agent, they would pay still less to a company's man, for they knew nothing whatever about the Honourable East India Company.

"The Government of Egypt" in 1798 might be variously described; it depended on the point of view. Egypt was an integral part of the Turkish Empire; the authority of the Sultan being exercised by a nominee from Constantinople bearing the title of Pacha of Egypt. Abubakr Pacha was in office at the date of the French invasion; he was removable at the Sultan's pleasure. He depended upon his body-guard—a considerable force—of Janissaries. There was, however, another armed force within the borders of Egypt, nominated by the Sultan, and set up for the express purpose of counterbalancing the authority of the Pacha, lest distance from Constantinople should tempt him to aim at independence. This was a Circassian militia officered by twenty-four Circassian Beys who owned, as well as commanded, their soldiery. This armed force went by the name of the Mamelukes. We can now follow clearly the tangled threads of power in Egypt. The ruler was the Sultan, but he never visited the country. He was represented by the Pacha, Abubakr, but the Pacha was powerless without his Janissaries. The Janissaries were overawed by the Mamelukes, who numbered 10,000 fighting men. They were supposed to represent the Sultan's interests as opposed to the Pacha's. In effect they represented their own interests and those of their leaders and masters, the Beys, of whom two were prominent. We must remember their names and qualities : Murad the Brave, and Ibrahim the Crafty. Both were rich. The Mameluke troopers were magnificent men, gorgeously attired ; haughty, overbearing and tyrannical, the real masters of Egypt. Bonaparte marked them for such, and duly set forth in all his proclamations that he had come to deliver the land from their oppression. The rest of the population of Egypt numbered two millions and a half, and consisted of Copts (the aborigines) ; Arabs, the conquerors of the Copts, who may be roughly described as Arabs of the cities (wealthy merchants), Arabs of the countryside (the Fellahin), and Arabs of the desert (the Bedouin) ; there remained the Turks, the conquerors of the Arabs. The large and important European colony of to-day had at that time no existence; the Beys paid no attention to English trade, a sure sign that it was insignificant. This did not deter Bonaparte from denouncing the Beys as the patrons of English trade to the detriment of Frenchmen, thus inflaming the minds of his soldiery by associating the Mamelukes with the enemy, against whom the expedition was directed.

Prior to the expedition of 1798 England and France were officially and privately as ignorant of Egypt, and as indifferent to her concerns, as both countries are to-day of Thibet. England remained ignorant, but not materially misled. France, however, was soon inundated with inspired accounts of the Land of Promise to which her sons had sailed. If the army of Egypt had been about to enter Antioch and dwell in the laurel-groves of Orontes, there might have been some excuse for the language that French writers permitted themselves. As it was, however, the first week's experience of Egypt reduced the whole army, from pleasure-loving Murat down to the most Spartan private, to a state of mutinous fury. Their conviction of their desperate position, and the prestige of Bonaparte's name sufficed to make them a fighting force : unwilling but effective. All hopes of colonization had disappeared before Desaix reached Dámanhúr.

On July 2, 1798, the day after the capture of Alexandria, Bonaparte interviewed the principal inhabitants. He repeated in his address the substance of the proclamation that he had caused to be drawn up and printed at sea between Malta and Alexandria. This proclamation had been distributed by the hands of Musulmán slaves rescued from the galleys of Malta. It set forth Bonaparte's personal friendship with the Sultan, whose enemies, the Mamelukes, he had come to destroy. It boasted of the overthrow of the Pope and the Knights of St. John, and claimed the gratitude of Islam for the destruction of a brotherhood professedly hostile to the faith of Muhammad. The proclamation and the address made a good effect, although Bonaparte greatly overrated it. Kléber was appointed to the command at Alexandria, but the Turkish governor was joined with him nominally. The Turk's name was Sherif Sayyad Muhammad, commonly known as al Karáim ; his local knowledge and influence was of course far beyond any that Kléber could hope to acquire. He held both to throw into the scale at the right moment, but he judged unfortunately in the sequel.

The advance on Cairo was now begun. Desaix left Alexandria at five o'clock on the evening of July 3, 1798, and marched on Dámanhúr. Kléber's command under Dugua started by way of Rosetta. Desaix's march is not hard to narrate. Sand, sun, heat apoplexy, the thirst of the desert mocked by mirage ; starvation, madness, a fury of despair ending in frequent suicides, ophthalmia and blindness—these were the furies that tracked the invading army, starting as it did without food, without water, without a reconnaissance or a map. The cry went up, "We are betrayed ;" Murat flung his

cap on the ground and trampled on it before his command ;
Lannes followed suit. " There go our executioners," growled
the rank and file as an officer cantered by. " They have sacrificed us," groaned the panting wretches who staggered out of
the ranks to die in delirium, or to await in consciousness a
lance-thrust from the hovering Bedouin. A march that was
terrible for the division of Desaix was yet more terrible for the
division of Reynier that followed it along the same route.
Desaix's men had drunk the wells dry. For Reynier's men
the problem was, therefore, to march four days through the
desert with nothing to drink and only biscuit to eat. The
fear of the Bedouin kept the men in, but discipline disappeared ; and if they had been attacked, sheer physical
exhaustion would ·have ensured their annihilation. Apparently, contempt for foot-soldiers prevented the Mameluke
cavalry from concerning themselves with the troops crawling
slowly towards Cairo through the desert. There was also
some doubt as to the intentions of the French. Abubakr
Pacha felt it to be his duty to resist them in any case ; but he
would have performed his duty reluctantly. If Bonaparte
was really advancing to overthrow the Mameluke despotism,
Abubakr would have sympathized with him ; but he could
have done no more than sympathize, for he was himself in the
power of the Mamelukes. If Bonaparte was invading Egypt
as a Turkish province, it was his duty to call out the Mamelukes for the defence of the country. He consulted Ibrahim
Bey, who consulted a French friend, who announced that
Bonaparte was on his way through Egypt to attack the
English in India. This was good news for Abubakr who was
now relieved of any necessity of giving orders to the Mamelukes—orders which they were not accustomed to receive.
He was about to offer help to the French when the noise of
cannon was heard. Ibrahim packed up his jewels and
prepared to flee into Syria ; Murad prepared for battle. In
spite of the facilities afforded him by the roaming habits of
the Bedouin, Murad was imperfectly informed as to the
fighting strength of the French, their numbers or their order

of battle. One thing he knew: that they were foot-soldiers; and as his own force was accustomed to look on infantry as mere marauders and camp-followers he had no doubt of his power to ride down and destroy any number of Frenchmen. As to their intention Murad did not trouble himself. The presence of the strangers must in some way menace his authority, and that was reasoning enough for the man of action.

The desert march ended at Dámanhúr, where the French arrived on July 7, 1798. Bonaparte left Alexandria at five o'clock of the same evening and reached Dámanhúr at sunrise on the 8th. Two days later Reynier and Vial left for the south, and Desaix followed on the 11th, leaving Bon at Dámanhúr. At Ráhmánieh they reached the Nile, and halted to await the arrival of the fleet and of the division which had marched by Rosetta. On the 12th the flotilla came up with the army, and on the 13th there was a severe action by land and water, in which the French suffered heavily; a charge by Murad's cavalry was repulsed with slaughter, and the French moved slowly on. A new plague visited the invaders in the shape of dysentery, the result of eager feasting on melons. But, in spite of every difficulty, the morning of July 20, 1798, saw the French army drawn up in a square in sight of the Pyramids, Desaix on the right, Kléber on the left and Bonaparte himself commanding in the centre. There had been from the first but one chance for the Mamelukes. If they had consented to bear some of the hardships which were the daily portion of the Bedouin, and had incessantly harried the French by day and night between the 7th and the 14th of July, this one week of irregular warfare would have saved them from the necessity of fighting the battle of the Pryamids. By the evening of July 20 the rank and file of the French army were amusing themselves with what they humorously called "la pêche aux Mamelukes." The magnificent cavalry that had ridden out of Cairo in the morning had disappeared as a fighting force. Murad fled to Upper Egypt with a handful of men; Ibrahim burnt his ships to prevent pursuit, and carried his harem and his treasures to Belbeys on the way to Syria. The rest of the fighting force of the morning were corpses floating down the Nile, or being fished out of back-waters by enterprising French privates, and despoiled of their shawls, their chains, their turbans and their gorgeous weapons. Bonaparte slept in Murad's palace at Gizeh, Egypt was at his feet.

On July 25, 1798, Bonaparte made his public entry into Cairo, and proceeded to take measures for the consolidation of his conquest. As regards religion he proclaimed full toleration for Muhammadanism. On being pressed he did not show any marked repugnance to conversion, but stipulated that he and his men should have time allowed them for studying the dogmas of the Prophet. On finding them satisfactory he promised to build a magnificent mosque—say in two years' time. Administrative affairs were entrusted to a Divan under the superintendence of Monge. The finances were directed by Poussielgue. Very little money was found at Cairo; so the treasures of Malta were sent for from Alexandria to be coined. The fate of Bonaparte's messenger to Kléber on this occasion was the first warning of coming trouble: the messenger was waylaid and murdered at the village of Alkam, and the village in consequence was burnt to the ground.

Bonaparte was justified in considering his conquest completed and his ultimate success assured. There were more troops under his command than England thought necessary, in after years, for the defence of country with a hundred times the population of Egypt. His civilian staff included the pick of the intellect of France, instead of the ne'er-do-weels and hobbledehoys with whom England had officered India for the past half-century. He might now justifiably turn his attention to his secret instructions.

The failure of France in Egypt is usually accounted for by references to the military and naval disasters that befell the armed forces of the Republic, by the battle of the Nile, Abercromby's expedition, and Baird's march through the desert. These are sufficient reasons; but there are others weightier even than these. It is obvious that although five-and-twenty thousand men are an adequate garrison for a country like Egypt, provided that there is every confidence in the Government, ten times that number will not suffice, if the invaders have not their heart in the conquest. Far from feeling the eager interest in their work that alone holds the promise of lasting success, the French troops—if their feelings may be judged by the letters of their officers—were profoundly depressed. A home-sick boy of sixteen, dreaming in the heat of a warehouse at Fort St. George, of the cool rectory garden or cathedral close in the old country, could not have written home more dejectedly than these conquerors of the East. "You cannot imagine the horrors of our march," wrote Admiral Perrée to Brueys. "This wretched dog-hole, Cairo, with its population of lazy ragamuffins," wrote Damas to Kléber. "We have all been completely taken in about the enterprise that has been cried up as such a noble one." "This country is nothing like what it was said to be." "Alexandria is unimaginably dreary, sordid and unhealthy." There were not lacking soldiers to whom even the military achievements of the army brought no sense of exultation. "This expedition and its victories will not seem so wonderful when people come to understand what kind of enemies we have had to contend with, how feebly we have been opposed, and how totally lacking they are in enterprise;" "This country is no good for us;" "Cairo is a hateful place;" "The army has had no comforts since we came to Egypt"—such letters as these are to be expected from boys newly landed in the East. Very considerable men have at first written in even more profound self-abandonment and despair —Charles Theophilus Metcalfe for example; but these are the letters of grave soldiers, veterans all, men responsible for the future of the country, men upon whom it depended that Cairo should cease to be a "villasse horrible," and become a stately city. If there had been any chance of Bonaparte's ultimate success, there would have been signs of it in the expressed resolve of his companions-in-arms to do away with the squalor of Alexandria, and to make the population of Cairo something more than "canaille paresseuse." But an incurable disease had fallen upon the army of Egypt from the first moment when the sands were sighted—a loathing for the country, for the work before them, for the very name of their new conquest; an unconquerable longing to see once more the coasts of France, and a hatred, mingled with a horror, of the people of Egypt, a state of mind out of which no good work ever yet came or ever will come.

Berthier, the chief of the general staff, was the most flagrantly home-sick of all. He was at this date forty-four years old, and his infatuation for Signora Visconti was common gossip with the army. When the staff made a picnic to the Pyramids, Bonaparte banteringly urged Berthier to make the ascent, on the chance of seeing or hearing something of Signora Visconti at the top. When the Syrian expedition was planned, Berthier had finally made up his mind to return to Italy, but the strongly-expressed disappointment of Bonaparte kept him at his post, and Louis Bonaparte went in his place; although one would have expected the head of the army to be overjoyed at the prospect of getting rid of a subordinate who daily bemoaned the miserable fate that brought him to Egypt. Berthier took no interest whatever in the work of the army.

If this was the temper of the chief of the staff, it was hardly to be expected that the younger members of the expedition would do better. Menou, governor of Rosetta, Berthier's only senior, was forty-eight. He married a beautiful Muhammadan girl, daughter of a bath-proprietor, and settled comfortably down to the life of a conforming Musulmán; not even the peremptory orders of Bonaparte could drag him from the harem. Friant, who was forty, was the only other important soldier who had passed his first youth. Bonaparte himself was twenty-nine, Desaix thirty, Murat twenty-seven, Davoust twenty-nine, Marmont and Savary (the one governor of Alexandria, the other Desaix's aide-de-camp) were only twenty-four.

It was with a staff and an army in this condition of moral decrepitude that Bonaparte prepared to subjugate the East. Lower Egypt had acquiesced passively in the passage of the French, and the General concluded that his work was done. But acquiescence was one thing : submission another. Village after village revolted, submitted to a punitive expedition, and revolted again. The authority of France in Lower Egypt was confined to the cantonments of the army; and Bedouin ravaged the open country up to the gates of Cairo. The situation was an exact parallel to the position of England in Morocco, a century and a quarter earlier. Enough force had been displayed to alarm and irritate the natives, not enough to terrorize them. One centre of disturbance was discovered in the person of the Governor of Alexandria, Sherif Sayyad Muhammad, commonly called Elkoraïm. He was arrested and sent on board the flagship *L'Orient*, and was later on executed as a traitor at Cairo.

Between the first suspicion of Sayyad Muhammad's treachery and his execution, there occurred the battle of the Nile, no account of which will be attempted here. Five combats took place at or near Aboukir, between July 1798 and March, 1801. Of these the three principal were the first battle of Aboukir, fought on August 1, 1798, between Nelson and Brueys ; the second battle of Aboukir, fought on August 25, 1799, between Bonaparte and the Grand Vizier ; the third battle of Aboukir, fought on March 21, 1801, between Sir Ralph Abercromby and General Menou, who had succeeded to the command-in-chief in Egypt after the assassination of Kléber. Of these the second and third are also described as the battle of Alexandria ; the first is also called the battle of the Nile. There was, in addition, the fighting before Alexandria on July 1, 1798, before Bonaparte entered the city, and there was another battle between Abercromby and the French on March 8, 1801. In these pages they are distinguished as follows—

    1. Capture of Alexandria, July 1, 1798.
    2. Battle of the Nile, August 1, 1798.

3. Battle of Aboukir, July 25, 1799.
4. Second battle of Aboukir, March 8, 1801.
5. Battle of Alexandria, March 21, 1801.

On August 1, 1798, the French fleet was destroyed. It was an event of the first magnitude, but the result is alone historically important, and the competence or incompetence of Nelson is of little consequence; the effect would have been the same if Nelson had never been born, and the French fleet had been wrecked in a storm like the fleet of Cloudesley Shovel. Accustomed as we are, however, to look on the battle of the Nile as a considerable naval exploit, it is perhaps instructive to recall the very different manner in which it was spoken of by the French.

According to French opinion it was undeniably a misfortune, but much overrated by the English as an achievement. The defeat of the French was not owing to superior ability on the part of the English, but to the decrees of fate. In fact, it was owing to the incompetence of Villeneuve that the English were not utterly destroyed. It appears to be admitted that the French captains were not all faultless, although the men performed prodigies; but it is impossible to award anything but blame to Nelson. The English Admiral, after showing his incompetence in the pursuit of the French, made an unjustifiable attack upon their anchored fleet. Afraid as he was of being deprived of his command, his reckless ambition impelled him to a step which can only be described as a forlorn hope.

When Suffren beat Hughes, the English were contented to conclude that Hughes was a good man, but Suffren was a better; it must be left to sailors to decide whether the more analytical methods of writing history adopted by the French are preferable to such blunt conclusions.

It is noteworthy that in matters of food and surgical appliances the French expeditionary forces had much to complain of. In our own days questions are asked in Parliament if anæsthetics are not available after a battle fought a thousand miles deep in the desert. A century ago Nelson did his best in the same direction. Remembering the sick shudder that he felt when the cold steel touched his own maimed arm at Santa Cruz, he directed the surgeons to have pails of hot water ready before the action, so that the knife might be dipped and brought to the temperature of the body before amputation was performed. There was no one to think for the French private in Egypt.

Bonaparte reproved Brueys for his anxiety as to the feeding of the sailors under his command; but although the General could not say enough of the fertility of Egypt, the fleet went into action half-starved. There was no money for the troops, who were badly clothed and had no shoes. In the course of Belliard's retreat from Assouan, there was practically no treatment of the wounded. A certain amount of ignorance of tropical diseases may no doubt be excused; for all doctors were at that time mystified as to the sources of enteric fevers, ophthalmia and the plague. But the shocking condition of the hospital at Rosetta is inexcusable. Small wonder that the private soldier, seeing so many well-cared for *savants* attached to the expedition, and apparently doing no work, came to have a hatred of civilians that expressed itself in a flood of taunts and insults.

Bonaparte's hopes were founded on the ancient achievements of Alexander and the modern achievements of England. As regards his modern model his feeling, his very natural feeling, was, "What the English can do, cannot the French do better?" To which the reply, surely, was— "Granted; but not twice as well, still less one hundred times as well." For Bonaparte was attempting to bring about in one year a state of things which had taken England a whole century to establish. But if the quasi-commercial operations which the example of England imposed upon him drew his attention at first, his most cherished model was Alexander the Great. The ambition to imitate Alexander was no doubt stimulated by the result of his engagements with the Mamelukes. The triumph of discipline over numbers appeared to be as marked in his own case as in that of

Alexander, and to promise the same results. But there were several important differences, which he either decided to ignore or really under-estimated. In Alexander's case there was no enemy of his own grade of civilization already established in Asia; vigilant, active and in some respects superior to himself. Then the climate of Asia must have changed considerably in the course of the last twenty-three centuries, or else the Greeks must have been more nearly of Asiatic temper than we are accustomed to allow. Bonaparte appears to have made insufficient allowance for the hampering effect of a hostile Europe behind him; and finally a great difference between the conditions of 1798 A.D. and 330 B.C. had been wrought by the rise of Islam. Alexander married a Bactrian princess; one has only to recall the ridiculous effect of Menou's behaviour at Rosetta to realize how completely the possibility of smoothing difficulties over by a fusion of conquerors and conquered had been taken away by the presence of Islam. When Darius fell there was no reason why his subjects should not obey his conqueror. But, apart from the fact that none of Bonaparte's victories were as decisive as those of Alexander, there was every reason why good Muhammadans should continue to resist the infidel. In effect, there was no cessation of repressive and punitive measures in Egypt from the first moment when the Egyptians realized that Bonaparte's visit was intended to be permanent.

Early in August 1798, Bonaparte undertook the conduct of an armed force that was directed against Ibrahim Bey, whose Mamelukes were still hovering on the north-eastern frontier of Egypt. At Salahieh, on the edge of the desert, after a smart cavalry engagement, Ibrahim was defeated, and driven into Syria. Leclerc, Beauharnais, Arrighi, Duroc and Murat were all present, and Caffarelli, in spite of his wooden leg, wrought like a common trooper. So far as fighting was concerned, there could hardly have been a band of men who better deserved an empire than the army of Egypt. But it was attempting the impossible. Hardly had the victory of

Salahieh cleared Lower Egypt from the Mamelukes, whom Bonaparte persisted in describing as his only enemies, when the little garrison of Mansurah was isolated and destroyed by Bedouin. Everywhere throughout the Delta small parties of Frenchmen, whether foraging or engaged in sketching and exploring, were cut off and murdered. Kléber at Alexandria, hemmed in between the Bedouin and the English, was in evil case. To add to his miseries, the plague broke out : and in the midst of this disorganization, Poussielgue was innocently complaining that his financial work was hampered by the absence of capital in the country.

Nothing daunted, Bonaparte proceeded with his work of peaceful organization, and on August 20 founded the Institut de l'Egypte at Cairo, with Monge as president and himself as vice-president. A month later the *fête* of the foundation of the French Republic was celebrated in Cairo with much pomp and festivity. The Crescent was interlaced with the Cap of Liberty, and the Koran with the Rights of Man. But this attempt to force antagonistic principles and institutions into alliance only brought into greater prominence the difficulties of the situation. Murad Bey declined the invitation of the French, and persisted in calling himself the faithful subject of the Sultan. "Urgent private affairs" were pleaded with increasing frequency by the officers of the French army. The General had to reprimand the surgeons for granting sick-leave certificates in inconvenient profusion. Hardly a month after the interlacing of the Crescent with the Cap of Liberty had proclaimed the alliance of hopelessly antagonistic principles, a fierce revolt broke out in Cairo, the causes of which form a curious comment on the credit claimed for Bonaparte for his enlightened dealings with the Egyptians. The causes of discontent were—Firstly, the precautions taken to deal with the plague ; we can well imagine that the measures of one hundred years ago must have been fruitful in irritating incidents. Secondly, the laws relating to the use of stamps for official documents and to the registration of titles to landed estate. Thirdly, the anger of the Moslems at

the diminution of their own privileges. Fourthly, their fury
at seeing Copts and Jews treated as their equals. Fifthly,
and as a direct consequence of the two last-named sources of
irritation, the efforts of a committee of Moslems who met in
the Mosque El Azhar to organize resistance to the odious
oppression of a government based on the Rights of Man.
Sixthly, the ferment produced by the report that the Turks
had landed at Alexandria, that Murad had marched on Gizeh
(where Desaix was camped) and Ibrahim on Belbeys, in order
to support them. Seventhly, the firman of the Sultan himself
denouncing Bonaparte and all his works. Why this firman
had been so long delayed, is not easy to decide. Mr. Morier,
private secretary to Lord Elgin, British Ambassador at
Constantinople, who was possessed of great natural acuteness,
and had ample opportunities for acquiring sound information,
says that sheer astonishment on the part of the Porte, and
nothing else, accounts for the delay. But, however that may
be, the firman, when it did appear, destroyed Bonaparte's claim
to be acting in the name of the Sultan, and produced an immediate effect on the minds of the population of Cairo.
Eighthly, the Cadastral Survey of Egypt. This was a harmless measure. It was undertaken by M. Testevuida, a man
of good disposition. He had previously surveyed Corsica, and
was now sixty-three years of age. In the coming tumult he
lost his life, his scientific labours having worn the appearance
(to the Egyptians) of one knows not what witchcraft and
wickedness. Ninthly, the direct exhortations to revolt
delivered by the Imáms. There seems to be no doubt that
they took advantage of the French ignorance of the native
language, and openly proclaimed a Jahád on the above-
named eight grounds.

The revolt took place on October 21, 1798. It was marked
by the usual sanguinary incidents of street-fighting, by the
deaths of Shulkowski (an aide-de-camp of Bonaparte's), and of
the head of the Cadastral Survey. Caffarelli's house, one
among many, was looted, and the scientific instruments found
there were smashed. The instruments were a severe loss :

they had been sent out from France for use in cutting the Suez Canal. The revolt was unsupported in the provinces, and its chief result was the fortification of Cairo—or rather surrounding Cairo with fortresses, to control the capital in future. Salahieh, Alexandria, and all points of vantage in Lower Egypt were fortified at the same time, and a purely military government was decreed. Very remarkable claims have been put forward for Bonaparte, claims which the revolt of October 21, 1798, and the grievances that produced it, utterly discredit. " Ferme et bienveillant à la fois, il entrait dans les mœurs du pays, et cherchait à les tempérer par un mélange progressif de notre civilization." This may have been what Bonaparte was attempting to do, but it is only with considerable reserves that we can admit that his measures were successful. Let us take this claim again. " Si quelques faibles rayons des lumières européennes se sont depuis fait jour au milieu des ténèbres de l'Orient, c'est à Bonaparte seul, à sa patiente intervention, qu'il faut rapporter la cause première de ce progrès." Surely of all the adjectives available, " patient " is the last to apply to Napoleon Bonaparte. One would rather conclude that the Egyptian expedition, in any case a work of incalculable difficulty, was made into a hopeless failure by the precipitation with which Bonaparte insisted on forcing through the most radical changes, and yoking together the most antagonistic principles.

Towards the end of 1798 the revolt in Cairo was suppressed, and the country-side terrorized by the exploits of an irregular police force, which did not always discriminate between the peaceful cultivators and the marauders who were their plague. Ibrahim had been thrown into Syria, the strong places of the Delta fortified, Cairo itself encircled with forts, and Desaix despatched to Upper Egypt to deal with Murad. Bonaparte now turned his attention to cutting the Suez Canal. On November 2, 1798, Beauharnais and Bon were despatched to occupy Suez in force : they reached their destination on the 8th ; Bonaparte followed on December 24, and entered Suez on the 26th. He was accompanied by Monge, Berthollet,

Bourrienne, Berthier, Caffarelli and Gantheaume, and addressed himself to the third of the articles comprised in his secret instructions. The vagueness of his published instructions has been alluded to; the objects of the Egyptian expedition were set forth more fully in the three following secret articles: " Article I.—The General-in-Chief of the army of the East will seize on Egypt. Article II.—He will drive the English from all their possessions in the East, and above all destroy their entrepôts in the Red Sea. Article III.—He will have the Isthmus of Suez cut through." " Suez était pour lui le chiffre connu d'une équation gigantesque; il voulait par une création merveilleuse, creuser sur cette terre égyptienne le tombeau du commerce anglais." But to break the line of communication was no longer the simple matter that it had been in the days when Portugal had aroused the jealousy of Venice. It had been comparatively easy to storm Ormuz, especially in the days of the Portuguese captivity; it was quite another matter to dig the Suez Canal. Nevertheless, long and difficult although the task appeared to be, the menace was none the less seriously taken in England. " The possession of Egypt will be a measure indispensable to the preservation of the British dominions in the East Indies." The loss of part of the scientific instruments destroyed by the rioters in Caffarelli's house during the revolt of the preceding month delayed the surveying operations; the loss of the rest in the *Patriote* reduced Bonaparte's work in this direction to the order that the Canal should be cut. He issued these orders on December 29, 1798, and left Suez on the 30th. On January 3, 1799, he was at Belbeys, and thence he rode on a tour of inspection through Tel-el-Kebir to Ismailia. This was the extent of the first Napoleon's dealings with the Suez Canal. Surveying work was commenced, in accordance with his orders, on January 20, 1799, and carried as far as Belbeys. Here it stopped: for on February 9, 1799, the surveying party entered Cairo and found that no more military escorts would be available for scientific purposes, and without an armed guard the survey could not proceed. In the last three weeks

L

the expedition had taken a new turn—the Syrian campaign had been started—and on the day when the Suez Canal survey party entered Cairo, Reynier was already before El Arish.

While he was at Suez Bonaparte wrote two letters, one to the Sultan of Mysore, and the other, a covering letter, to the Imám of Muscat. The covering letter ran as follows—" I write you this letter to inform you of the arrival of the French army in Egypt. As you have always been friendly, you must be convinced of our desire to protect all the merchant vessels you may send to Suez. I also beg you will forward the enclosed letter to Tippoo Sahib by the first opportunity." The letter to Tippoo Sahib ran as follows—" You have already been informed of my arrival on the shores of the Red Sea, with a numerous and invincible army, animated with the desire of delivering you from the iron yoke of England. I hasten to inform you of my desire to receive news with regard to the political position in which you find yourself placed. I even desire you will send to Suez some competent person who enjoys your confidence and with whom I can confer." In April 1799 the city of Seringapatam was taken by storm, and the Sultan's correspondence with the French was printed and published. In this volume this letter is described as having been sent through the Sherif of Mecca, a provincial governor, who remained on good terms with the French, for the excellent reason that as soon as the French had seized Upper Egypt, Mecca must of necessity procure corn from the French or starve.

In January 1799, Bonaparte found himself in this position—his communications with India were confined for the present to the letters just quoted ; and his naval force in the Red Sea consisted of four despatch-boats, and one corvette not yet ready for sea. Clearly, therefore, there was no possibility of invading India. He might advance up the Nile, but Desaix was already at work on that task. For the rest, the Nile was unexplored beyond Assouan ; there were no reputed great kingdoms beyond, which it would be either glorious or profitable to subjugate, and the further he sailed up the Nile the

further he left France behind. If he moved north he came to
the sea, and the line of British cruisers. There remained two
alternatives. He might remain in Cairo and watch his measures
take root and bear fruit. Most generals would have been
contented to do so. Already the armed occupation of Suez
had won over one tribe of Bedouin. The Suez Arabs, finding
that they could no longer live by levying blackmail on the
Mecca caravans, were glad to enter the French service in
various capacities. Already, as we have seen, one outpost to
India, and an important one—Mecca itself—had been virtually
conquered by the occupation of Upper Egypt. There were
even signs that the hostility of the population of the capital
was diminishing, so that Mr. Morier, writing a year later, in
defence of the convention of El Arish, went so far as to say
that unless a speedy peace took the French out of Egypt,
we might have a nation to deal with instead of an army.
These were hopeful signs, and any general who was contented
to sit still and watch for the day when France recovered the
command of the sea would have been justified ; but he would
not have been Napoleon Bonaparte. It is not, however,
entirely just to say that Syria was Bonaparte's only way out
of a trap ; for later, when the lines were drawn yet closer
around him, Bonaparte escaped by the simple process of
shipping on board a French vessel of war, and evading the
British cruisers. But it was undoubtedly the only course open
to him, if he wished for immediate glory and success. It was
a course that was covered by his vague and grandiose instructions, and it was foretold in England that he would take the
road east through Bagdad, enter Afghanistan from the north-
west, and join hands with Zemán Sháh. His own prophetic
musings took him back to Paris by way of Constantinople and
Vienna. As a matter of history he was driven back from Acre,
to which fatal siege we shall shortly have to trace the footsteps
of the army of the East. In the Egypt of January 1799 the
generals of France were distributed as follows—Desaix was
in Girgeh, struggling with insurrection ; Kléber, Governor of
Alexandria, defending the city walls from the cannonade of

Sir Sidney Smith, kept down as best he could the Bedouin who threatened him from the open country, and the angry citizens who were outraged at his plague regulations; Menou was in Rosetta, waiting for the day when the repetition of the *Kalimah* would smooth the way for "Abdallah Menou" to enter on that agreeable harem existence which he afterwards declined to quit, even for the governor-generalship of Palestine; Dugua in Mansurah and Vial in Damietta kept order inside their cantonments, outside was anarchy. The physical condition of the soldiery had improved; the plague was yielding to strict sanitary measures; dysentery, with more regular food, better water and no exposure, had practically disappeared; the wounded were recovering with rest and care; but ophthalmia was raging. An addition to the army in the shape of a camel corps, five hundred strong, had taken place at the end of 1798; its principal achievement was a temporary tranquillization of the Fayum. The new stress and burden that was laid upon the army which had already suffered so much was the act of Bonaparte. No orders, or indeed communications of any kind, had reached him from the Directory since he had landed in Egypt, and there was a strong feeling that peace would be more advantageous to France than any fresh conquests. Desaix, to whom the task had been confided of chasing Murad and his Mamelukes, and if possible taking them captive, had by now discovered that any idea of "conquest" was illusory. An armed occupation of territory was possible; but "conquest" in the sense of tracts of country reduced to obedience, and content to settle down with the French as their rulers, was as far off as ever. The army under his command had suffered severely from ophthalmia, while it was camped in the Fayum; otherwise it was in fairly good condition for its long march. The lack of money was the worst evil that had to be endured; but this was common to the field forces of Egypt everywhere. Two hundred thousand pounds, actually all that the plunder of Malta produced, were soon absorbed by the daily needs of an army of thirty thousand men. The hoards which Eastern rulers think it their duty to accumulate were

few in Egypt. Murad's hoard was inconsiderable. Ibrahim
had carried his treasures with him to Syria. Contributions
were levied in Alexandria, and on leading citizens of Cairo.
But these are precarious resources ; and it was with a war-chest
scantily supplied, and with troops poorly fed and badly clothed
that Desaix prepared to chase Murad and his Mamelukes.
He broke up camp at Medinet el Fayum on November 2,
1798, and arrived at Siút, on December 25. On the 29th, he
reached Girgeh, the capital of Upper Egypt, and waited for
the flotilla to join him. Murad had, so far, retreated before
the French, but not without employing the authority which he
still wielded, even in adversity, in stirring up discord in the
rear of the French army. His information and local knowledge
were, of course, incomparably better than Desaix's, and as
the result of his efforts, Siút rose in rebellion directly the
French had passed through, and moved south to Girgeh.
Davoust was detached to deal with the insurrection, which he
did without much difficulty. Whenever the French and the
Arabs were face to face the issue was certain. Throughout
January 1799, Desaix continued his rapid march, Murad
always retreating and repeating the manœuvre of Siút. On
January 24, Desaix reached Denderah, on the 26th Thebes ;
but it was not until the 28th that he first came in touch with
his nimble enemy. On the last-named date, Desaix entered
Esneh and learnt that Murad had only left the village twelve
hours before the arrival of the French. Beyond Edfu the belt of
cultivated land narrowed, and the army found itself in great
straits, but on February 2, 1799, Desaix entered Assouan,
and found some evidence that his pursuit had at last become
too hot for Murad to retreat in comfort; the Bey was compelled to abandon some boats. Neither the French nor the
Arabs passed Assouan. The pursued broke and scattered
to right and left ; Elfy Bey threw himself into the desert
eastwards ; Murad, with Hassan and Solyman Bey, plunged
into the Gizm Halfa. Here ended Desaix's march into Upper
Egypt. It was devoid of results, except scientific results. It
may be doubted whether the Beys were ever seriously pressed

or distressed, except in so far as men of middle-age who have long enjoyed leisure and authority must naturally be distressed by a series of hurried marches. But their campaign could hardly have been better designed. The French were led a thousand miles away from their base, and they had great difficulty in recovering their communication with Lower Egypt. Pursuit of the Beys was out of the question when they had taken to the desert, and Desaix, recognizing his powerlessness, left Belliard at Assouan with a light corps and betook himself to Esneh, where he arrived on February 4, 1799. The decision seems to have been taken in haste, but there was nothing to be gained by waiting at Assouan, and the reason for leaving even a part of the force there is not apparent. As an exhibition of courage and endurance, the march on Assouan was no doubt remarkable, especially with an army so badly provided. As a military effort it had two results. Firstly, it was very encouraging to the enemy who had now learnt his strength and how to use it. The desert was for the Arabs a natural fortress, where (as the army began to realize) they could not be pursued, and where they miraculously recuperated their forces. Secondly, the Nile campaign profoundly depressed and angered the army. To the rank and file it appeared as if they were doomed to suffer, in order that the learned men who accompanied them might make sketches in safety. From the first the presence of *savants* had been looked on with suspicion. The march on Assouan aggravated the distrust and ill-feeling with which they were regarded. Devoid as it was of military interest, it was of immense value scientifically. The *savants* were busy all day long ; the soldiers regarded them with anger and disgust ; this, then, was what they had been brought to Egypt for. It was in vain that Desaix sought every opportunity of coming to close quarters with the Arabs ; to fight a decisive battle would have been to throw away all the advantages that knowledge of the country gave the Arabs. To make the French draw up in battle array, and then to dissolve their own ranks and disappear in the desert, was, for the Arabs, an effective and harassing manœuvre.

It is only by reading between the lines of the French narrative, that one can form any idea of the extent to which Murad by his disappointing tactics succeeded in exasperating and fatiguing the French. Once, in the course of his retreat down the Nile, Desaix contrived to edge a body of Arabs into a corner between Kosseir and Keneh, but the result was not all that he anticipated ; there was a smart skirmish, and the French under Davoust were roughly handled. The garrison left in Upper Egypt was cantoned in Esneh : it numbered five hundred men, and was, naturally, attacked so soon as the main body of the French had moved north. This bold measure on the part of the Arabs was ascribed by the French to sheer desperation, at the prospect of being driven into the desert to starve. But Egypt above Assouan is not a Sahara ; it might be an uncomfortable place of residence for Frenchmen, but assuredly not for Arabs. The attack on Esneh was repulsed ; and the commandant moved on Assouan, which was thus (May 16, 1799) a second time occupied by the French. But to say that this measure consummated the con-quest of Upper Egypt is to mislead. It is evident that at no time between November 1798, when Desaix started, and May 1799, when he was again in Lower Egypt, did the French control any part of the country not in actual occupation of their forces. Murad had evaded all attempts to capture him, and was now settled down in an oasis three days' ride from Siút. Desaix, as unwearied in pursuit as Murad in flight, was preparing to pursue him with a camel corps when the news reached him that the English fleet had been sighted off Kosseir. Belliard and Donzelot were despatched in haste to occupy the port ; and Desaix himself followed, entering Kosseir on May 29, 1799. Henceforth we cease to be con-cerned with Upper Egypt. Scientifically the expedition was valuable ; it was not in any sense a conquest, and Murad was not captured. It is perhaps instructive to recall that while Desaix was hunting Murad through Upper Egypt another young man of the same age was hunting an equally nimble fugitive through the Southern Mahratta country. The

difference between the two performances was that Desaix did
not catch Murad, but Wellesley did catch and hang Dhúndia
Wág. It is not of course implied that Wellesley was in any
sense the superior of Desaix ; but only that, having been
some years in the country (an advantage which was not
enjoyed by Desaix) Wellesley had learnt local conditions. He
saw that it would be useless to campaign in a red soil country
as one might campaign in Kent ; that it would be useless to
announce moral victories, enormous losses on the part of the
enemy, and the fearful hardships to which he was put by the
pursuit. The only reasoning that would be understood would
be the patent and undeniable fact of Dhúndia's execution.
To attain that end he must ride as light as Dhúndia, fare as
hardly and march faster. Desaix campaigned in Egypt as if
he were in Italy ; and Murad rode round him, doubled in his
tracks and mocked at his pursuit. Nor did Desaix mend
matters by burning the whole village of Abu Girgeh to the
ground with men, women and children alive in their houses,
as a lesson to the country-side. This measure did not
terrorize ; it only infuriated.

The presence of the English in the Red Sea was a serious
matter. It appears that they had even presumed to reconnoitre Kosseir and estimate the chances of a landing in force.
But the presence of Desaix was decisive. After fortifying the
seaport, and leaving a considerable armed force behind him
(a singular measure when he had recently been " welcomed
with enthusiasm ") to keep the population down, he withdrew
to assist at the triumphal entry into Cairo of the victorious
army of Syria. His immediate head-quarters were at Siút ;
for a few days there was peace.

All this time we are to remember that England was in full
possession of the despatches and letters written from Egypt,
while the Directory had no news of its army, and the General
had no instructions from France. The intercepted letters
were published by order, at which the French were highly
indignant. England, in fact, seems to have borne herself in
this crisis in the history of France with more than her usual

malignity. Not only, as we have seen, did she presume to reconnoitre Kosseir, but she had a representative at Constantinople whose influence was used against the interests of France, although it surely does not require the presence of English agents to explain the anxiety and alarm with which the Sultan regarded the invasion of one of his provinces by a powerful French army. Nor did English perfidy stop at hostile suggestions. It appears that the unfavourable attitude of the famous Pacha of Acre, Djezzar, was to be ascribed to their hostile intrigues. Djezzar had the assurance to occupy El Arish, the news of which outrage reached Bonaparte at Suez, and contributed to his irritation. El Arish is on the coast, midway between Gaza and Port Said, and was afterwards described by Bonaparte as one of the keys of Egypt. Ibrahim Bey had been allowed to dwell in Syria instead of being handed over to the French. When Bonaparte sent to demand his immediate expulsion, Djezzar beheaded the envoy. It was time to chastise so insolent a neighbour, and Bonaparte assembled 13,000 men for that purpose. At this time Talleyrand was supposed to be the Ambassador of the French Republic at Constantinople, but he had refrained from taking up his appointment, and France was represented by Ruffin, her *chargé d'affaires*. Ruffin was seized and thrown into prison; war against France was declared; Hajji Abubakr, the Pacha of Egypt, was deposed by firman, but not disgraced; his place was taken by Hajji Abdallah, with whom Murad and Ibrahim were ordered to co-operate; the English fleet kept open the communication by sea, and (as we shall see) rendered very material services to the Porte. So long before this date as August 22, 1798, Bonaparte had written to the Grand Vizier a despatch intended to soothe his anxieties, and on the eve of his invasion of Syria he sent Beauchamp to lay before the Porte his intentions, which were to punish England and the Mamelukes, and to hinder the projected partition of Turkey between Germany and Russia, to all of which overtures no response was made, except to throw Beauchamp into prison in company with Ruffin.

The military objects of the campaign were to broaden the area of desert lying between Egypt and Syria, and to lay Jaffa, St. Jean d'Acre and El Arish in ruins. In writing to the Directory, Bonaparte stated that he was about to assure the conquest of Egypt, to come to a definite understanding with the Sultan, and by assuring the friendship of the Syrians, to cut off the supplies hitherto obtained by England. Privately he was heard to mutter, " But three steps hence— Constantinople, Vienna, Paris." He was evidently misled by the apathy with which the Porte had witnessed his Egyptian campaign. He imagined that he had to deal with an enemy devoid of spirit or energy, and that the capture of Constantinople would prove to be as easy as the capture of Malta. The arrest and imprisonment of Ruffin and Beauchamp was followed by the arrest and imprisonment of the French Consuls at Algiers, Tunis, and Tripoli; the Emperor of Morocco acted in the same way as the Sultan. Thus from end to end of the Mediterranean, Islam was aroused against the French. The slow-moving Turkish mind had at last grasped the situation; the Egyptian expedition was not a raid, not a mere passage of armed forces through Turkish territory to India; it stood out as a menace of the most serious nature. Bonaparte was now to experience how much of dogged resistance he had roused, and how little impression had been made by his protests of friendliness to the Sultan, and affection for his religion.

The new province of the Turkish Empire that he was about to invade contained a population of approximately two millions and a quarter—as nearly as can be estimated, the same as the population of Egypt. It was divided into five Pachaliks—those of Aleppo, Tripoli, Acre, Damascus, and the loosely defined territory of " Palestine." The most famous of all these Turkish governors was the Pacha of Acre—Ahmad Pacha. Ahmad was a man of obscure origin, reputed to be a Bosnian, and has passed into history by his self-conferred nickname, " the Butcher." The mis-spelt Arabic equivalent has become too well known in its French form for transliteration to be useful, and Ahmad will probably continue to be described in history as "Djezzar." There was no Turkish field force in existence that could be called an army, or that could move or manœuvre with anything approaching order. The so-called troops were accompanied by at least an equal number of camp-followers from whom they were hardly distinguishable. But they fought well behind defences, and the hills of Samaria and the Lebanon were full of armed tribesmen, who seriously embarrassed the French. Bonaparte fought but one pitched battle in the course of the Syrian campaign, but he went fully prepared for any resistance, and was accompanied by Reynier, Lannes, Bon, Kléber, and Caffarelli. Murat, as usual, commanded the cavalry. The mounted force numbered 900 men. The artillery was sent by sea, a fatal blunder. Of the generals left in Egypt, Dugua commanded in Cairo and Lower Egypt; Destaing commanded the city garrison of Cairo, and Menou stayed in command at Rosetta. The command of Alexandria was conferred on Marmont. Salahieh was the starting-point for Reynier. This was the town on the edge of the Syrian desert where Ibrahim Bey had fought his last engagement with the invading army in the early days of the expedition. Kléber was directed to move by Kattieh. Reynier left Belbeys on January 23, 1799, and plunged into the desert. Apparently the French had not profited by experience. Reynier's march was a repetition of Desaix's march on Dámanhúr, excepting that by this time the anger of the soldiers had fixed on the civilians who accompanied the army as the sources of their trials. The *savants* were incessantly gibed at, and the civilian element was indiscriminately described as *savants*. The learned people who were sent into Syria were mounted as well as might be, while the soldiers marched on foot. This was another reason for detesting them. When no enemy was in sight, they rode as far as might be from the toiling soldiery, thus avoiding the dust of the march and the stream of abuse and coarse pleasantry with which they were greeted. The soldiers had their revenge when Bedouin appeared, and civilians were ordered inside the square. "Donkeys and civilians inside," translated the mockers, and so the ill-matched companions moved forward together till the danger was past. On February 9 the siege of El Arish was commenced. Ibrahim Aga commanded the garrison, which was 2000 strong. There was a stout resistance, and continual house-to-house fighting. The French mined, the garrison countermined; the village was carried, but the fort held out. Meantime the plague broke out again at Alexandria, and Marmont had to report some ominous movements of the English fleet. Marmont's news was so persistently gloomy that he earned the nickname "General Bad-Luck." On February 10, 1799, Bonaparte left Cairo and arrived at El Arish a week later. His itinerary was :—

10th, Belbeys	14th, Kattieh
11th, Karaim	15th, Bir el Abd
12th, Salahieh	16th, Messoudiah
13th, in the desert	17th, El Arish

The distances in this famous march were :—

Cairo to Salahieh	.	23 leagues
Salahieh to Kattieh	.	16 ,,
Kattieh to El Arish	.	24 ,,
El Arish to Gaza .	.	17 ,,
Gaza to Jaffa	.	18 ,,
Jaffa to Acre	.	23 ,,
Total		121 leagues,

of which seventy only were through cultivated country, the rest being through the desert. The siege of El Arish took three days, that of Jaffa four; there were eleven days' halt. The average day's march was therefore about twenty miles. On the 20th the defenders capitulated; they were allowed to retire to Bagdad, and the French left a garrison in the citadel and established a hospital there. They marched on Gaza, which surrendered at a summons, and here the army halted for two days, while Bonaparte issued a proclamation stating that he was a friend of the Syrians and had come to deliver them from the tyranny of Djezzar. On March 1, Bonaparte arrived at Ashdod, and on March 4 and 5, trenches were opened for the siege of Jaffa. On the 7th the garrison was summoned to surrender, a summons to which Abu Sahib (following perhaps Djezzar's example) responded by beheading the messenger. The assault was ordered, although the breaches were not complete. But Bon discovered a secret passage into Jaffa from the northern side. The fortress was captured and given up to sack and pillage for two days. Three thousand prisoners who had surrendered on terms were shot in cold blood. It is unnecessary to enter here on the question whether or no Bonaparte's explanation was sound or not. The point for consideration is that the large number of prisoners slaughtered on this occasion made concealment impossible. It became known throughout the confines of Syria that no terms were observed by the French. It was thus inevitable that they should meet with the most desperate resistance everywhere. It was no part of Bonaparte's plans to appear as a second Tamerlane, but the events of Jaffa showed him in the light of an exterminating conqueror, and aroused a hostility that he had not the force to overcome.

Jaffa was erected into the capital of Palestine, and Menou was appointed governor-general, with Robinu under him as governor of the citadel, and Gloutier as administrator-general of finances. The only one of these appointments which was operative was Robinu's, as Menou declined to leave his harem, and Palestine yielded no finances for Gloutier to administer. The plague now broke out, and the hillmen of Samaria began to be troublesome. Kléber was directed to repel them, but was ordered not to pursue them to the hills. Kléber was minded to try a hill campaign, but Bonaparte with indubitable wisdom forbade him; he had no mind, he said, to meet the fate of Crassus. Acre must be taken before anything else was considered. Unfortunately for his plans, other people obtruded themselves on his notice. Abdallah, the newly-appointed Pacha of Egypt, had approached from the north to effect a juncture with the Samaritans, and Bonaparte was compelled to detach Bon and Lannes to prevent him. They succeeded in doing so, but Lannes was carried away by his hot temper. He allowed himself a pursuit into the hills, and was entrapped and severely punished. " J'ai voulu chatier cette canaille," he said to Bonaparte, who replied, " Nous ne sommes pas en mesure de faire de pareilles bravades." On March 17, 1799, Kléber occupied Haifa, and from Mount Carmel he sighted the English fleet. It was Sir Sidney Smith in the *Tigre*. The next day Bonaparte drove in Djezzar's skirmishers and settled down to the siege of Acre.

The whole of his siege train (including four twenty-four-pounders) had been captured by Sir Sidney Smith as it was being conveyed from Damietta to Haifa. The guns were landed and placed in position on the ramparts of Acre. Djezzar, a man of high capacity in spite of his monstrous reputation for cruelty, gladly availed himself of the services of European officers. There were many such at hand. Sir Sidney Smith placed the resources of his fleet at the Pacha's disposal, and the process of transforming the antiquated fortifications of Acre into a modern fortress was put in hand only too thoroughly. The directing engineer was Phélipeaux, a French emigrant, who had been trained at Brienne with Bonaparte. He died before the siege was over, but lived long enough to know that his genius for defence had baffled the attack of his old school-fellow. The two ships under Sir Sidney Smith's orders held the command of the sea, and their fire reached considerably to the north and south of Acre, and searched the entrenchments of the besiegers whenever they ventured too near to the coast. The *Thésée* was posted to the north, the *Tigre*—Smith's ship—to the south. Acre was thus completely invested on the east, partly invested on the north and south ; communication of the garrison with the open country was impossible, but it was out of the question for the French to attempt to interfere with communications to the west with the fleet. The country-side was hostile to the French, and the Pacha Abdallah was advancing from the north. The problem therefore was to reduce Acre as soon as possible. Its capture would curb the hostility of the hillmen, and would give the French the comfort of living in a walled town instead of in open trenches, although, as the town would immediately be shelled by the British fleet, it is clear that the greatest possible success attainable by the French would only be a transition from one awkward situation to another. A long siege was out of the question, as Bonaparte's resources were limited, and those of Djezzar were practically unlimited, so long as Smith held the command of the sea. Nor was the temper of the army suitable for detailed operations of war. The soldiers were cast down at the first reverse. Outside the entrenchments there were hostile movements in all directions calling the French away from the siege. Vial was detached to settle one such "incident" at Tyre; Murat was ordered to march on Sidon; Junot was detached for duty in Samaria. Murat, having pacified Sidon, marched on Lake Gennesareth, while Junot occupied Nazareth and came into collision with the troops of the Pacha of Damascus. The combat received the name of the battle of Mount Tabor, or Loubyeh. In the meantime, things had not gone well with the French at Acre. On April 1, 1799, a mine was exploded, but the assault failed; a sortie was repulsed, but not without heavy loss. The news of Junot's encounter with the Pacha was too much like the news of a defeat to satisfy Bonaparte, who detached Kléber to support him. There was another fight at Cana of Galilee on April 10, but the open country was not settled until Bonaparte himself joined the field-army. On April 16, in the battle of Esdraëlon, he overthrew the Pacha of Damascus at the head of 35,000 Turkish troops. The French force numbered 4000; they claimed to have slain 6000 of the enemy, and admitted a loss of 200 on their own side. On April 17 Bonaparte slept at Nazareth, and on the 20th he returned to the trenches before Acre. On the 24th a new mine was exploded, but both the assaults that followed the explosion were repulsed. There was renewed fighting in the trenches on the 25th, with heavy losses on both sides, and on the 27th Caffarelli died after long sufferings—he had lost an arm three weeks before, and never completely rallied from the effects of the amputation. Bonaparte was as resolute as ever, and declared that when only four Frenchmen were left he would lead those four to the last assault. But this temper found no echo in the ranks, and among the officers only Murat's spirits were proof against so much misfortune. Murat dressed every morning as carefully as if he had been in Paris, performed his duties with the smiling and confident air of a man in the full tide of success ; braved every danger, it is hardly necessary to say, as if it did not exist ; dined at as great length as possible every day, and was continually surrounded with young fellows, whom he infected with his invincible spirits and his extravagant foppishness. The Turks adored him. Djezzar afterwards confessed that in one of the assaults he had an excellent opportunity of killing Murat—Múrád as he called him—but could not find it in his heart to slay a soldier so gallant, so handsome, and so brave. Bonaparte the Turks never regarded but as an evil genius, in spite of his quotations from the Koran ; but Murat was the incarnation of all the qualities that they most admired, and never was Murat more admirable than at Acre.

But the French were doomed. On May 6, 1799, another assault was delivered and repulsed, and the next day the whole army was made to see and understand what a hopeless task was before them. Thirty sail were sighted. Could these be the French fleet come to drive away the English, and complete the blockade ? The disappointment was bitter, and went home to every Frenchman in the army when the vessels proved to be Turkish men-of-war bringing provisions and reinforcements for the garrison and its English allies. Throughout May 7 and 8 there was continuous fighting for twenty-five hours, and one assault was only repulsed with the assistance of a party of marines and blue-jackets landed in haste from the *Tigre*. The last assault was delivered on May 10, when Arrighi and Eugène Beauharnais were both wounded, and the last sortie was repulsed on the 16th. The plague had in the meantime broken out, and despatches had arrived from Egypt bringing the news of Desaix's retreat down the Nile. It was now that Bonaparte heard for the first time the news of the English approach up the Red Sea. Kosseir was not yet menaced, and by the time that Desaix entered that port Bonaparte was in the desert halfway between Khan Younès and El Arish on his march back to Cairo. But the news was alarming. Moreover, an insurrection had broken out in the Fayúm with a Mahdi at its head. On May 20, 1799, at nine in the evening, the siege of Acre was raised and the army of Syria began its retreat. The siege had lasted sixty-two days, and the French had delivered fourteen assaults and repulsed twenty-six sorties. The garrison did not care to pursue their foes, and the march having been begun at night, the army was well out of reach of the English guns by break of day. It reached Cæsarea on the 22nd, and Jaffa on the 24th. Jaffa is famous for two sinister events, in the campaign of 1799; the first is the massacre of prisoners on March 7, and the second the poisoning of the sick on May 24. Both atrocities have been affirmed and denied with equal vehemence, the latter especially. Desgenettes, the head of the medical staff, flung the charge in Bonaparte's face in full council at Cairo, when the commander-in-chief was indulging in his usual sneers at doctors. Desgenettes may be supposed to have been acquainted with the facts, but the story is no longer one that interests the world. It is an incident of the Syrian campaign; the campaign was but an incidental development of the Egyptian campaign. Egypt was but an episode in the career of Napoleon, and so vastly has the political horizon widened in the last hundred years that Napoleon himself is but an episode.

There was a three days' halt at Jaffa, and on May 28 the army resumed its march. On June 1 it entered the desert; camped at El Arish on the 2nd, and at Kattieh, on the 5th and 6th. Here Menou, the Governor-General of Palestine, joined the army to make his peace with the commander-in-chief; he succeeded in doing so, and in spite of his incompetence, was never afterwards unemployed. On June 8 and 9 the army

M

re-entered Egypt by Salahieh, and on the 14th it made a triumphal entry into Cairo. The declared objects of the campaign having been to chastise Djezzar, to drive the Turks north and to cut off the sources of supplies for the British fleet, one must conclude that the campaign was a failure; insomuch as Djezzar was still Pacha of Acre, the English were as well provided as ever, and the Turks were shortly about to invade Egypt. The French appear to have found consolation in the fact that they retired before incidental obstacles, and not before a more skilful enemy. The "incidental obstacles" were, roughly speaking, Phelipeaux and Sidney Smith.

The most important event that had transpired in Lower Egypt during the absence of Bonaparte had been the rise of the Mahdi in the Fayúm. At the end of April 1799 the garrison of Dámanhúr was attacked and put to the sword. Marmont, who promptly moved out to deal with the new enemy, was defeated, and by May 3 the Mahdi's army numbered fifteen thousand foot and four thousand horses. The Mahdi fixed his head-quarters at Dámanhúr, and there awaited the arrival of the Beys from Upper Egypt. A sanguinary engagement at Senhar had resulted in the virtual defeat of the French, who retreated, however, in good order: a general rendezvous was ordered for Rahmanieh with the object of concentrating forces sufficient to break up the rebellion at once. On May 10 the army marched from Rahmanieh to Dámanhúr: the entire population of the latter town was exterminated, and a few days later the Mahdi was slain, and his army dispersed. Murad and Hassan Beys, with whom the rising had been concerted, retired to Upper Egypt. The Fayúm was once more reported to be " calm."

Since the battle of the Pyramids Murad had not met Bonaparte; but two months after the death of the Mahdi, he moved from Lake Natrún on Gizeh, and Bonaparte marched out to meet him. True to the tactics which he had so successfully employed, Murad at once took to the desert, and the French marched back to Gizeh. Here on July 14, 1799, Bonaparte received from Marmont, Governor of Alexandria, a despatch reporting that on the 11th he had sighted seventy-five sail heading for Alexandria. This was the Turkish fleet conveying Mustafa Pacha and the army of invasion. Here at least was an enemy who would not run away, and Bonaparte hastened to meet him.

The triumphal entry of the army of Syria into Cairo had been made on June 14. On July 14, Bonaparte heard news of the coming enemy; with the exception of his attempt to seize Murad, he had in the meantime taken part in no field operations. His time was spent in endeavouring to persuade the chiefs of the Musulmán population that he was himself a Mahdi. He laid very great stress on gaining the confidence of the native population; but he seems to have misunderstood their mental attitude. Having decided, from the outset of his campaign, to conciliate their religious prejudices, he announced his impending conversion to Islam from the Grand Mosque of Cairo immediately after his departure for Syria. As well as assuming the character of a Mahdi, he incessantly sounded the college of Mollahs as to what proof they would accept of his divine mission. It was a question of building a mighty mosque, the dimensions of which were laid down in the Hadisa. But he could make no way. He saw that the Egyptians were easily conquered, he saw that they made patient subjects, but he did not realize how much of genuine and stubborn religious conviction was enshrined in this soft human clay. Nor, to the end of his days, did he grasp the fact that conquest in the style of Alexander had been made impossible by the rise of Islam; and that the utmost Europeans could hope to attain in Egypt would be to live side by side with the Muhammadans, tolerated and perhaps advised by them; but only accepted as rulers so long as Egypt lacked the physical force to expel them.

Bonaparte's career in the East had now entered on its last stage. He himself appears to have kept his illusions till the last. "This battle decides the fate of the world," he muttered before the day of Aboukir. "Say, rather the fate of the army, general," said Murat, who overheard him. Murat's "limited intelligence" was a phrase often on Bonaparte's lips in after life, and was a constant excuse for disobliging him. It dated from Murat's refusal to take the campaigns of Egypt and Syria as anything but good pieces of fighting. Menou, on the other hand, who humoured the commander-in-chief, was allowed to idle away his time without being rebuked, and never, afterwards, wanted employment.

The Turkish army of invasion numbered eighteen thousand men. To meet it Bonaparte ordered Murat to collect the cavalry and concentrate at Gizeh. Reynier was to provision the forts guarding the eastern frontier, and keep watch over Ibrahim Bey. On Friant was laid the double duty of keeping Murad in sight, and at the same time keeping in touch with Dugua, the Governor of Lower Egypt ; he might as well have been ordered to keep in sight of Murad, and yet never to quit the banks of the Nile. Lannes and Rampon were ordered to Ráhmánieh, and Kléber was to move on Rosetta. On July 19, 1799, Bonaparte arrived at Ráhmánieh, and six days later he fought the battle of Aboukir. Nine thousand Turks perished ; the French admitted a loss of ten killed and forty wounded. In the proclamation after the battle, dated on the anniversary of the battle of the Nile, Bonaparte informed the victorious army that they had reconquered the French possessions in India. "Nous venons de reconquérir aujourd'hui nos établissements aux Indes." No doubt this extravagant statement was well-chosen for the effect it was designed to produce. Probably, also, it reflected Bonaparte's desire, if not his actual conviction. As a matter of fact, the battle of Aboukir was a barren victory, devoid of lasting results, even in Egypt itself, and outside Egypt it produced no effect whatever. Three weeks later, at ten o'clock in the evening of August 22, 1799, Bonaparte embarked on the frigate *La Muîron*. He was accompanied by Bourrienne, Gantheaume, Berthollet, Monge, and Denon. Marmont, Murat, Lannes, Duroc, Beauharnais, and Bessières, sailed at the same time in the *Carrère*. The command-in-chief in Egypt was entrusted to Kléber. They had a safe, but tedious passage of thirty-five days, and anchored in Ajaccio on September 26. On October 7 they set sail again, and landed at Fréjus on October 9. On November 1 the Directory was overthrown, and Bonaparte became First Consul. His career in the East was at an end ; and at this point we may profitably endeavour to compose the contradictory accounts as to what was the effect of his Egyptian campaign, and what was the state of the country when he left it.

In reading French accounts of the Egyptian and Syrian campaigns we are struck with one fact ; the French never appear to lose any men. We read of hundreds, and perhaps thousands, of the enemy being exterminated, cut to pieces, burnt alive, or driven into the desert to perish of starvation. The French occasionally admit having lost one man killed and thirteen wounded. At the battle of Aboukir, where fourteen thousand of the enemy were killed or captured, the French lost ten killed and forty wounded. England has recently campaigned so much in Egypt that we know that these statements cannot be exact. British losses at Tel-el-Kebir and Omdurman were not very severe; compared with those of the enemy they may even be said to have been insignificant. But they did amount to several hundreds of men ; and that in spite of a superiority of armament on our part, very greatly in excess of any superiority enjoyed by the French over the Bedouin. Then it is easy to read, on the face of the narrative of Desaix's march down the Nile, that he was compelled to fight every stage of his retreat from Assouan to Siút. It is clear that Murad outmanœuvred him, drew him away from his base, and then, taking to the desert, doubled on his tracks and closed with Desaix at every favourable opportunity. There was, further, the siege of Acre, with fourteen unsuccessful assaults delivered and twenty-six sorties repulsed. There was the march through the Syrian desert and the retreat over the same ground. The plague haunted the army everywhere, and there must have been considerable losses from other sicknesses. Kléber, who was furious at the desertion of Bonaparte, stated that by the time that he took over the command half the army had perished. Bonaparte challenged this statement as soon as possible. This was not immediately, as Kléber's despatch was captured by the English cruisers. Bonaparte was highly indignant at Kléber's description of the army of Egypt, and replied to it by saying that the army when he left Egypt was still twenty-four thousand strong. It had originally been thirty thousand strong, and three thousand men, in addition, had been saved from the fleet. The losses of the army he estimated at 4500, leaving twenty-four thousand men all in excellent fighting condition.

Three years after the date of this controversy, the chief of the medical staff of the army of Egypt published his official report. He estimates the total losses of the army for one year at 8915, as follows :—

Killed in battle	3614
Died of wounds	854
Accidental deaths	290
Deaths from sickness other than the plague	2468
Deaths from the plague	1689
	8915

This is almost exactly twice the loss estimated by the commander-in-chief.

The report is a smooth official production, admirably lucid on all points not affecting the army, and fearlessly mendacious upon occasion. Thus it does not hesitate to affirm that no army was ever better looked after than the army of the East ; and it even claims what not even the best appointed service of our own time can claim—that the medical staff kept the field hospitals up to the standard of garrison hospitals, even in moments of defeat. It is not now even pretended that these statements are true ; so that the admission of a loss of 9000 men coming from this source may be taken to represent the minimum. But we are still a long way from Kléber's figure of 15,000 men destroyed. Let us consider the report of Poussielgue. Poussielgue was a civilian, but being entrusted with the clothing and financing of the army, his report as director-general of finances must needs carry weight. His view is even more gloomy than that of Kléber. He points out that there were twenty-two garrisons to be kept up, and that when these places were manned there remained only 5000 or perhaps 6000 men fit to take the field. The certainty of the plague recurring periodically weighed heavily on the spirits of the army, and every man from the generals downwards was sighing for the moment when he might return to France. The commander-in-chief had carried away the best generals with him, and of the others five useful general officers had been slain, viz. Dommartin, Bon, Caffarelli, Dupuy, and Rambault. In effect the army was without clothes, arms, or ammunition. Having completed his report on the actual state of the army, Poussielgue proceeded to some political reflections which must have been gall and wormwood to Bonaparte, so great is the contrast between their cold common sense and his own fervid imaginings. He condemned the expedition as premature, and insisted strongly on the need of a powerful navy ; the unalterable physical conditions of the country demanded it. To attempt to hold Egypt without a navy was merely to invite Russia or England to expel the French, to establish themselves there on the pretext of freeing the country, and to shut out France altogether. So long ago, and so clearly did Poussielgue indicate the three Mediterranean Powers whose rivalries were destined to meet in Egypt. The mass of controversy on the subject of Egypt that has accumulated in the course of the last century is, after all, little more than commentary on those few sentences of Poussielgue's despatch to the Directory. The efforts of England have been always misunderstood ; those of Russia naturally circumscribed by the want, until quite recent times, of a navy. But the outlines of the struggle were laid down a century ago.

We have travelled some distance from the question how many men did the army of the East lose between August 1798 and August 1799. The weightiness of Poussielgue's despatch must serve as an excuse. It is desirable, perhaps, to pay some attention to what may appear a question of administrative detail, because much has been made to depend therefrom. The partisans of Bonaparte condemn Kléber as a faint-hearted man, taking always the gloomiest view, and they maintain that Kléber's lack of spirit, after he assumed the command-in-chief, is to be held accountable for the failure of the campaign. All that he had to do (so Bonaparte's partisans, following their chief, allege) was to build on the foundation laid by Bonaparte. Kléber's champions, on the other hand, affirm that no foundations had been laid at all ; that all attempts to conciliate the people had failed ; that the army was in rags, and its pay in arrears, the whole country seething with discontent, and the very army itself on the verge of mutiny ; that the English held the Red Sea and the Mediterranean ; that the boasted loss of 9000 men inflicted on the army of Mustafa Pacha, was really nothing when the resources of the Turkish Empire were considered ; and that all that was left for Kléber to do was to conclude a peace as little disastrous as possible. It is to be noted that the army of Egypt was composed partly of the army of the Rhine, whose hero was Kléber, and partly of the army of Italy, whose hero was Bonaparte. Bonaparte's adherents dwelt much on this.

The dispute between Kléber and Bonaparte is only of value to us to-day, because we wish to find out what was the state of the army of Egypt when Kléber took over the command-in-chief. Both generals made reports : Kléber's was moderate, cool and reserved like his character, and he concluded by announcing his intention of concluding peace as soon as possible. If there is any truth in Bonaparte's reply, Kléber's report was a tissue of cowardly fabrications. The question of the losses endured by the French army has been chosen for examination here, and we find that on that cardinal point the mendacity was all on Bonaparte's side. One is the more inclined to believe Kléber because, as we shall find in his dealings with Sir Sidney Smith, he did not hesitate to brag, and brag mendaciously upon occasion, if he thought that he

could serve his country in that way. It was only to the
Directory that he told the truth. From the date of Bonaparte's departure, the only question that occupied the mind
of his successor was how soon could France return to the
status quo ante. The French were on the defensive, and even
defence had become difficult. Considering the inflated
language to which the army had become accustomed, it is to
Kléber's credit that he should have perceived the only course
that could now profit France. He alone appears to have
comprehended that the assault of France had welded together
the states of Islam in a common resentment, and that the
evacuation of Egypt would profit the Republic indirectly
even more than directly. The factors in the problem were :
in Egypt the irrepressible Murad Bey, in Syria the Grand
Vazir, at sea Sir Sidney Smith. Of these, the Grand Vazir
was a courtier, doing a soldier's work, and doing it badly.
Murad was an admirable guerilla chief with no pretensions to
being a regular soldier, still less a diplomatist. The situation
was in the hands of two men, Kléber and Sidney Smith—the
one a good soldier, the other a good sailor, and both of them
good negotiators.

The two chiefs esteemed each other, and both were sincerely
desirous of a peace ; neither expected to gain it without considerable concessions. October 30, 1799, stands out as a day
on which Kléber approached Sidney Smith with a series of
statements that one can hardly read with gravity. Sir Sidney
was given to understand that Kléber by no means considered
his position to be altered for the worse by the reverses of the
Republic and the fall of Seringapatam. His own resources in
men were, he said, abundant, for he could raise what native
levies he chose. We may note that it is precisely at this very
point that the French plan of action broke down. If either
Bonaparte or Kléber had really been able to raise twenty
thousand native auxiliaries the whole affair would have worn
a different complexion. Finally, Sir Sidney Smith was given
to understand that the manufactories of powder and ball
established in Egypt by the French furnished him with

abundant supplies of the munitions of war. This letter was written five days before the despatch to the Directory, in which the general explained that the French foundries and manufactories in Egypt were an almost total failure.

This may have been useful boasting in so far as Smith was concerned ; but its effect was somewhat discounted by some earlier letters. On August 17, 1799, Bonaparte had written to the Grand Vazir by the hand of Mustafa Pacha, captured at Aboukir, and had practically offered to come to terms. This letter, dated only three weeks after the battle, was highly significant. No reply was sent. Nevertheless, a month later, on September 16, Kléber himself wrote to the Grand Vazir to the same effect as Bonaparte. So much eagerness to come to terms was inconsistent with the confident tone of the communication addressed to Sir Sidney Smith. The terms offered by Kléber were the restoration of the *status quo* as the price of the French evacuation of Egypt. Specifically the Ionian Islands were to be restored to France, the English were to raise the blockade of Malta, and the French were to be allowed to disembark part of the army of Egypt in Malta without any definite understanding as to the length of their stay there. Practically, that the French were to retire, but to be given every facility for their next invasion. To the Grand Vazir Kléber pointed out that the French had had no object in their invasion of Egypt, except to alarm the English for their Indian trade and possessions, and to affirm the authority of the Sultan. All the advantages of the French campaign had accrued to Turkey, so Kléber maintained. The Grand Vazir's reply was haughty ; Kléber had over-stated his case ; and it was clear that the Grand Vazir would not recognize overtures made in such a temper. Sir Sidney Smith stepped in, and induced both sides to entrust the case to plenipotentiaries who might meet him and arrange terms. Kléber did not assent without writing the boastful letter before alluded to ; although it was notorious that the military difficulties were not his only difficulties. The religious irritation at the presence of the French increased daily ; and no taxes could be collected,

except by imprisoning leading citizens wherever the impositions were resisted, or, by the more summary process still, of
driving away the cattle. However, having put a good face on
affairs Kléber nominated Desaix and Poussielgue to represent
France, and negotiations were opened on December 21, 1799.
They proceeded slowly, and on January 5, 1800, they were
resumed at Gaza, whither the commander's flagship had been
transferred for the sake of convenience. The anchorage off
Damietta had severely tried Desaix, who suffered from sea-
sickness.

At Gaza the news reached the negotiators that the French
garrison of El Arish had been massacred, after it had agreed
to surrender with the honours of war. Bonaparte had described
El Arish as the key of Egypt. Kléber considered it a wretched
and insignificant post. Whatever its importance it was now
in the hands of the Turks; but Kléber admitted that the
massacre was the result of a misunderstanding, and allowed
the negotiations to proceed. On January 24, 1800, the Convention of El Arish was signed. It was ratified by Kléber on
the 28th, and both French and Turks proceeded to put it into
execution. The principal articles provided that the French
army should concentrate on Alexandria, Rosetta, and Aboukir,
there to await transport to convey them to France. Considerable sums were to be paid to the French to cover the expenses
incidental to the evacuation of the country. Various dates
were fixed for the withdrawal of the French troops from the
garrison towns. The Turkish authorities were to preserve order.
There were regulations as to passports; protection was provided for all who might have favoured the French during the
occupation of the country, and a three months' armistice was
proclaimed. Such were the principal provisions of the Treaty
of El Arish, the credit of concluding which must be awarded
to Sir Sidney Smith. He it was who assumed the direction
of the proceedings from the first, and by his tact and skill
made negotiations possible and fruitful. He had described
himself throughout as plenipotentiary, which he was not; and,
unfortunately, he was not authorized to conclude any treaty

whatever with the French. In pursuance of the terms of the convention, the Turkish army was pushed into Egypt until its outposts were within ten miles of Cairo. On February 18, 1800, Kléber and his plenipotentiaries returned to the capital, and three days later Desaix, Davoust, Miot, Savary and Rapp sailed for Europe. The preparations for removing the rest of the staff and the rank and file were in an advanced stage, when there was placed in Kléber's hands a letter from Admiral Keith, dated January 8, 1800. This letter, written in ignorance of Sir Sidney Smith's proceedings, set forth in dry and formal language that the Admiral could only regard the French in Egypt as prisoners of war. Sir Sidney Smith hastened to assure Kléber that his action would assuredly be recognized in the end ; but Kléber could have drawn only one conclusion from the unfortunate misunderstanding ; he could only conclude that he had been trapped ; that the English had deliberately deceived him, and were now preparing to force him to an ignominious capitulation. He therefore caused Keith's letter to be printed *in extenso* and published ; he added three lines at the foot, calling his troops to arms to resent the insult. The action was dramatic, the language fine and simple ; the army was roused from its inglorious lethargy, and the Turks were bidden to withdraw from Egypt.

There had been three parties to the convention of El Arish—the English, the French and the Turks. The English could do nothing except express their regret ; for Keith was Smith's superior officer. The Turks declined to budge from their advanced posts, alleging that the convention would assuredly be recognized. But Kléber was mortally offended ; and the army—with whom the evacuation had not been altogether popular—was only too anxious to fight. It was hardly to be expected that English expressions of regret would have much effect. Desaix and his companions had been captured at sea and made prisoners of war. It is true that they were subsequently released, and that it was explained to them that their arrest was the result of a misunderstanding. But Kléber had no evidence that our action was other than designedly perfidious. The English he could not touch : but the Turks, with whom (according to the only view possible for him to take) our treachery had been concerted, were in the field. On March 20, 1800, Kléber fought and gained the battle of Heliopolis. The Turks were utterly defeated. Donzelot, Belliard, Friant and Reynier shared the glories of the day with Klébcr. The pursuit was kept up, and in four days of head-long rout the Turks were driven from Egypt into Syria. By March 27, 1800, Kléber was back in Cairo, having gloriously re-asserted the honour of France. The Convention of El Arish had been approved by Lord Elgin, the British Ambassador at Constantinople. Lord Elgin arrived at his post on November 6, 1799, and wrote to Lord Mornington on the 17th, saying that he was authorized to regard the Embassy as a means of concerting plans to keep our Asiatic and Mediterranean politics in touch with one another. For the moment Egypt dominated the situation, and he did not see how we could get the French out of Egypt without an Indian expedition, " however honourable and brilliant our naval operations" might be. Consequently it was with great satisfaction that he watched the able diplomacy of Sir Sidney Smith. The commodore was justified in feeling that he had done a piece of good work. Writing to Mr. Dundas on February 7, 1800, he alluded, among other difficulties which he had surmounted, to " the trouble I have had in preventing the Turks losing all the advantages of the home-sick disposition of the French by a presumptuous confidence that their motley multitude was equal to driving them out." Writing to Lord Mornington the day before he prophesied that Kléber, of whom he had a high opinion, would be the Monk of the French Revolution. It was felt on all sides that the Convention of El Arish solved many a difficult problem. Then came its repudiation—unintentional, but none the less outspoken—by Admiral Keith, and all the disastrous consequences. We have been acquitted of treachery or perfidy towards Kléber ; but the blunder did its work at the time, and was to cost England three years of intense anxiety and many valuable lives.

The immediate effect of the repudiation of the convention was to strengthen the position of France in Egypt, not only by the military advantages gained at Heliopolis, but by the rapid development of the character of Kléber. The story of the French occupation of Egypt has three definite chapters falling, respectively, under the headings Bonaparte, Kléber, Menou. Menou was incompetence personified : Kléber was to Bonaparte what Bussy had been to Dupleix. The natives never trusted Bonaparte; but Kléber gained more influence in three months than Bonaparte had done in eighteen. Kléber did not pretend to be a Musulmán; he gained his influence by his inflexible uprightness of character. His military talents were more than sufficient for the demands upon them. Up to August 22, 1799, he had been a sceptical observer ; from that date until March 1860, he had been the unwilling and uninterested head of what he had always regarded as an impossible enterprise. The affront, as he deemed it, of Keith's rejection of the Convention of El Arish transformed him : he became the haughty resolute chief of an army—still fifteen thousand strong—of the veterans of France. He was undeniably the first man in Egypt, and though he thought poorly enough of the country he rose easily to the position of its ruler. Even the finances felt something of his influence, and looked less forlorn than before. A revolt at Cairo and the neighbourhood during Kléber's absence was the last flicker of discontent among the native populace. It was suppressed with vigour and not too much bloodshed. Murad Bey was recognized as tributary ruler of Upper Egypt, and this sensible measure relieved the strain upon the French treasury and their armed force in the field. Kléber, watchful in every direction, reserved the right to garrison Kosseir with French troops. Remembering the failure of Bonaparte's attempts to conciliate Egyptian sympathies, one is somewhat taken aback to find so shrewd an observer as Mr. Morier reporting that if we do not get the French out of Egypt soon, we shall have a nation to deal with, and not an army only. Surely if anything was clear in Bonaparte's enterprise, it was that the

French must always remain a mere garrison and that an amalgamation with the native population was impossible. But when we study the work of Kléber we perceive that, although amalgamation was as far off as ever, as far off as it is to-day, there was rapidly growing up a confidence in his integrity, capacity and military powers which, in the event of an invasion, would undoubtedly keep the Egyptians neutral, and might even allow of the enlistment of some native regiments. It is interesting to compare the views of Colonel Missett, British Consul-General in Alexandria, with those of Poussielgue, the Commissary-General of the French forces. Both agree that Turkish government is impossible, Egyptian self-government unlikely to be realized, and conquest from the desert not to be permitted. It is merely a question of who the foreign ruler ought to be : Poussielgue naturally puts France first, Missett equally naturally puts England first in the respect of Egyptians.

In June 1800 everything promised well for the French. Malta still held out, and was to hold out for another three months ; there was no rumour of an English expedition in any force ; the Turks had been scattered and driven out of the country. Under Kléber's rule the country was settling down and beginning to prosper. On June 14, 1800, Kléber was assassinated by a religious fanatic. It is impossible to over-estimate the effect of this tragedy. It occurred on the day of Marengo, a victory which cost France the life of Desaix ; so that one twenty-four hours saw Bonaparte relieved of the presence of two dangerous rivals, but also saw his only chance of success in Egypt fade away with the life of Kléber. Marengo laid Italy at his feet ; but for Bonaparte Italy was chiefly valuable as the high road to Egypt, and Egypt was now to be governed by General Abdallah Menou.

Menou had passed the preceding two years in his harem at Rosetta. While the comrades who were now submitted to his orders had been fighting from Assouan to Damascus, Menou had avoided all responsibility and all danger. He had been attached to the expedition because of his reputation

as a man of affairs ; and his adroit flattery of Bonaparte had
earned him forgiveness for his inglorious, not to say cowardly
behaviour. He was entirely incompetent, whether as a soldier
or an administrator. Nevertheless, although he might have
continued to lead unremarked the existence which he had
hitherto found agreeable, he eagerly seized the command-in-
chief which devolved upon him by seniority after the murder
of Kléber. Nor, apparently, was it by an oversight that
Menou was left in the chief command ; for on November 3,
1800, he received from Paris the formal confirmation of his
appointment. During these five months, June—November
1800, the situation may be described as follows : the English
were discredited diplomatically by the repudiation of the
Convention of El Arish. Their action could only be military,
and for military action they were not yet prepared. Besides,
they were wasting much energy, and employing a considerable
force in the Belleisle expedition which was intended to raise
the West of France in the Royalist cause. The Turks had
not recovered from the shattering blow of Heliopolis. The
Egyptians were still awed and quiescent under the effects
of Kléber's vigorous administration ; and the French were
condemned to inactivity by the total incompetence of their
commander.

In the meantime, Malta had surrendered to the English
(September 5, 1800) ; the Belleisle force had been forwarded
to Minorca and added to the army of Sir Ralph Abercromby.
With Minorca and Malta in English hands, and a powerful, if
not yet overwhelming English naval force in the Mediterranean, it was clear that the holders of Egypt would soon
have to fight for their position in that country. England in
1800 was a very different Power from the England who had
fled from Corsica in 1797. Small reinforcements had, indeed,
been forwarded to Egypt, but the French army received no
considerable addition to its numbers; and in the meantime the
English were drawing nearer. On November 2, 1800, Sir
Ralph Abercromby left Gibraltar and concentrated at Minorca.
Three weeks later he sailed from Minorca, and on December

29, 1800, the British fleet and convoys cast anchor off Marmorice in order to concert a campaign with the Turkish army. The position of 1798 was reversed ; the English moved their army confidently along the Mediterranean, while the French were unable to interfere with them.

To arrange a campaign in concert with the Turks was found to be an impossibility. With the important exception of courage they possessed no single military quality. Undisciplined when in cantonments they dissolved into a mere rabble when on the march, incapable of orderly movements, and destitute of any notion of keeping an appointment or even of decently providing their troops. Leaving the Turks, therefore, with the understanding that they were to advance through Syria and enter Egypt by El Arish, but in that counting much on their support, Sir Ralph Abercromby weighed anchor on February 22. He was prepared to attack the French with his own army, in conjunction if possible with that of Sir David Baird. Baird had started from Calcutta while Abercromby was between Minorca and Marmorice. When Abercromby appeared off Alexandria on March 1, 1801, Baird was nearing Bombay with the Anglo-Indian contingent on its way to Egypt.

If Kléber had been alive, or if Menou would have deferred to his highly capable subordinates, the danger of defeat would have been very considerable. Belliard and Reynier were not great soldiers ; but they had profited by their service under Bonaparte. They urged on the commander-in-chief that he could not possibly hope to improve on Bonaparte's strategy. They reminded him how the Turkish invasion in July 1799 had been repelled. To concentrate every man on the sea-coast, to prevent the enemy from landing, or to force him to attack a strong position—such was the appropriate plan for the French to follow. Menou would not give his consent. He left Belliard with half the army in Cairo, apparently for no other reason than that Cairo was the capital. He sent a small detachment to watch the Syrian frontier (which was not yet threatened), and he forwarded the remainder down the

N

Nile towards the sea-coast, under the command of General Friant.

For ten days after March 1, 1801, there was much anxiety in the English camp. After March 11, the English army had little to fear. It had lost some men, a thousand perhaps ; but it could afford that loss better than the French could afford their own losses, amounting to about one-half that number. The army was safely landed with stores and ammunition, and Menou now began to think of giving battle. He left Cairo on March 12, arrived at Alexandria a week later, and attacked Abercromby on the 21st.

The result of the battle of Alexandria was the defeat of the French. Menou has been blamed for the disaster ; while he, in turn, threw all the blame upon his subordinates. In England we have been accustomed to conclude that the victory was due to the superior ability of Abercromby, who was severely wounded in the battle and died the next day. Whether the defeat of the French was due to their being overmatched, or was merely an unintelligible decree of fate, the result was the same, and was serious for their cause. Menou shut himself up in Alexandria. There was no rising of the Egyptian population on either side. Hutchinson succeeded to Abercromby's command, and we have now to consider the almost bloodless campaign in which he succeeded in clearing Egypt of the French. Compared with Bonaparte's campaign, it was leisurely : and the French have no words of contempt strong enough for it. But it was successful, and it was not costly. Hutchinson reasoned that he was well-placed, drawing supplies from the fleet, and holding a strong position in the face of a defeated enemy. There was therefore no need for haste. Time was on his side : he was not called upon to expose his soldiers to dysentery, ophthalmia and heat apoplexy, in order to emulate the performances of a military genius. Even if he did nothing in his cantonments, Baird was drawing steadily nearer, and in time even the leisurely Grand Vazir might be counted on to have reached the Egyptian frontier. But Hutchinson was far from idle. He employed the time between March 21, and May 7, 1801, in flooding the space of ground known as Lake Mareotis to-day. The effect of this engineering operation was to cut off Alexandria from all possibility of communicating with Cairo. So far as a civilian can judge, Hutchinson, by calling in natural obstacles to do the work of another division, showed excellent common-sense. By May 7, Baird was only ten days from Kosseir, and the Grand Vazir had entered Egypt. Leaving six thousand men to hold Menou in Alexandria, Hutchinson now moved south. Belliard, completely isolated, had nevertheless pushed his outposts as far as Rahmanieh. His troops fell back before the advance of the English, who followed the Nile route, and were covered and supplied by the fleet. As the English moved with the utmost caution there were no surprises possible, and as they were abundantly supplied, there was no chance of their ceasing to move, however slowly they elected to march. It was not until June 23, 1801, that Belliard, with Baird distant only three days' march south from Cairo, and Hutchinson actually in Gizeh, prepared to capitulate. Generals Morand and Donzelot represented him. The Turks were a party to the surrender, and were represented by Isaak Bey and Osman Bey, who had succeeded to the leadership of the Mamelukes on the death of Murad the Brave two months earlier. General Hope represented Hutchinson. The terms granted were highly honourable, being, in fact, those of the Convention of El Arish. These terms had been proffered in good faith by Mr. Morier to Kléber after the battle of Heliopolis, but the offer had been rejected in scornful silence by the victorious general. They were now accepted by Belliard and the garrison of Cairo. Thirteen thousand six hundred soldiers of all arms and civilians were shipped on transports, and convoyed down the Nile. Three days after the surrender Baird reached Cairo. He had not fought a battle ; but his march through the desert was a remarkable piece of work, and no doubt the news of his advance accelerated the surrender of Belliard.

A curious feature of these almost forgotten campaigns is

the extraordinary amount of quarrelling that took place among the leaders on both sides. On the side of the French there was the fundamental difference between Bonaparte and Kléber, which told considerably throughout the campaign and burst out publicly in Kléber's bitter censure of the conduct of the campaign when he succeeded to the chief command. Bonaparte's violent reply made matters worse. On Kléber's assassination, the egregious Menou, the butt alike of officers and men, succeeded to the command and retained it, daily quarrelling with everybody with whom he came in contact. He openly reviled his subordinates, and ended by placing Reynier under arrest and sending him to France. On the side of the English there was the unfortunate repudiation of the Convention of El Arish. In this matter we have been entirely acquitted of treacherous conduct, but the effect produced at the time was that Keith publicly repudiated Smith. Abercromby, like Kléber, was universally respected, but on Abercromby's death Hutchinson succeeded to the command. The view here taken is that his operations were highly creditable to him. But at the time, and before his plan developed, he was as unpopular with his army as Menou was with the French army. There was, then, a violent quarrel between Baird, who claimed an independent command, and the commander-in-chief, and there were four changes in the chief command during two years—Abercromby, Hutchinson, Cavan and Stuart succeeding each other at short intervals.

To return to the course of the campaign after the fall of Cairo. There now remained only one effective French army in existence—that of Menou. Menou, shut up in Alexandria, issued bulletins announcing great victories and the slaughter of thousands of Englishmen. He vowed that he would be buried in the ruins of Alexandria if he were not victorious. There could be, however, but one result. Two months after the fall of Cairo it was clear that the defence could not be prolonged; a three days' armistice was granted, and the entire armed force of French capitulated on the terms of the Treaty of El Arish on August 31, 1801. Eleven thousand two hundred and thirteen soldiers of all arms and civilians were shipped to France, and the French occupation of Egypt was over. It had lasted three years and two months, from the capture of Alexandria on July 1, 1798, to the capitulation of Menou on August 31, 1801. What was to be the future government of Egypt? So far as England was concerned there could be no doubt and no hesitation; the country must return to the condition in which the French found it, and the English would have nothing more to say to its administration or its military affairs. But the course of the war had aroused a desire in the minds of Turkish statesmen to use the opportunity afforded by the serious damage wrought to the power of the Mamelukes, and to make the control of Constantinople over the vassal province more effective than it had hitherto been. This was natural, although the Mamelukes were still 3500 strong and were far better soldiers than the Turks. Some working arrangement might have been settled with English aid, if the English would have consented to take a share in the new settlement. But Hutchinson, Cavan and Stuart had but one object—to wash their hands of Egypt and all its affairs, restoring the *status quo* unaltered in every respect. The Turks, thereupon, it is disagreeable to remember, resorted to treachery and assassination. Several of the Beys were murdered and their bodies thrown into the sea. The rank and file were entrapped, and were no doubt destined to the same fate. But Hutchinson interfered, and fiercely ordered the Grand Vazir to have the bodies of the murdered chiefs recovered, under penalty of the English opening fire on his camp. The bodies were recovered by divers, and Hutchinson caused them to be buried with the military honours accorded to full generals. The Turkish fleet was ordered out of Egyptian waters, and as the result of this and other vigorous measures, the Mamelukes of Cairo and Alexandria joined hands and once more presented a compact armed force. These events bring us to November 1801, and in the meantime events in Europe had dictated a change of attitude on the part of the English in Egypt.

The preliminaries of the Peace of Amiens had been signed in London on October 1, 1801. On the 7th of the same month, a new treaty of peace and amity was concluded between France and Turkey, and four days later a young man whose name was to attain to a European celebrity within the next year—Colonel Sebastiani—left for Constantinople to seek the ratification of this treaty from the Sultan. If England was to retain a shred of influence with Turkey, it was evident that no time must be lost in repudiating the policy in which there had been such incidents as those just narrated. It was therefore made clear to the Mamelukes that England could no longer interfere in their favour. But the Mamelukes, when once they had been delivered from gaol and the certainty of assassination, were quite capable of taking care of themselves. They now formed, as of old, an armed body occupying Upper Egypt, while the Turks held Lower Egypt. On February 8, the Grand Vazir withdrew to Syria. On March 27, the Treaty of Amiens was signed, and on June 10, 1802, the Anglo-Indian regiments withdrew from Alexandria, embarking at Suez for India on July 6.

From the Turkish point of view the English had behaved very unreasonably. They professed the utmost respect for the authority of the Sultan, and yet they insisted on the restoration of the *status quo*, an essential feature of which was that the authority of the Sultan in Egypt was a nominal authority only. They declared that they only desired to stand clear of Egyptian embarrassments, and yet when the Sultan's generals proposed to solve those embarrassments in the only way in which they could be solved, viz. by treachery and assassination, they had most unreasonably stepped in and restored the perplexing situation of a viceroy with no power face to face with Mamelukes without authority. The first viceroy was Khosrou Muhammad Pacha; the title of viceroy had not previously been accorded to the rulers of Egypt.

The treaty of October 7, 1801, between France and Turkey had not been ratified. But insistence at Constantinople brought about the signature of another treaty equally effective. It was pointed out by the French agents that the friendship between England and Russia was ominous. The death of the Emperor Paul, six months before the signature of the preliminaries of peace, had deprived France of an ally. Before the Treaty of Amiens was signed, it was already clear that the former ally had become an enemy. An alliance between France and Turkey, to counterbalance the formidable Anglo-Russian understanding, was signed on June 25, 1802, and ratified on August 25, 1802. Promptly Stuart was directed to relax his unbending attitude towards the Mamelukes. He was not allowed to aid them, but he did not interfere on behalf of the viceroy, as he might have done, when they invaded Lower Egypt in the autumn of 1802. They gained a great victory at Dámanhúr, and minor victories wherever the Turkish troops encountered them. The year 1802 saw the authority of the Sultan decline every month. Had there been any constructive ability to be found in the Mameluke camp, the same year would have seen the establishment of an independent Egypt. But after gaining a victory the Mamelukes could think of nothing but plunder and pleasure.

While the Mamelukes were preparing their descent on Lower Egypt, Sebastiani started on his historic journey to Egypt and Syria. It lasted from September 16, 1802, when he sailed from Toulon, to November 16, 1802, when he cast anchor in Zante. His report on his travels is that of an intelligent observer, travelling with good introductions, and writing with a strong French bias. He was received with civility by General Stuart, who commanded what was left of the British army in Egypt. Stuart was somewhat mystified as to Sebastiani's position, but he paid the colonel every civility. The "strong French bias" is perfectly intelligible and pardonable ; but the distribution of large numbers of portraits of Bonaparte appears to be significant. The assurance that Sebastiani gave everywhere that Bonaparte would soon be back in Egypt, is a piece of conversation if Sebastiani were an ordinary traveller ; it is something more when we remember that he was clothed with diplomatic functions. To charge Stuart with conniving at assassination was outrageously rude. To write a report on the number of British troops remaining in Egypt—4430—lay, no doubt, within the limits of his instructions, but to estimate the number of Frenchmen necessary to defeat them—6000—is an act of hostile intent; and to publish the whole report to the world—which was done in the *Moniteur* of January 30, 1803—was a defiance.

The good faith of England in the negotiations of the next four months has been challenged. It has been affirmed that the breach of the Treaty of Amiens was brought about by the action of England. The manner in which the Foreign Office conducted the negotiations, as well as the irritating objections raised by England—these, and not the ambition of Bonaparte, must be held to be responsible for the renewal, after so short an interval, of the war between the two countries. Let us consider what actually occurred.

As early as June 1802 there were signs that France had no intention of allowing old sores to heal. This, at least, is the obvious explanation of the objections raised in that month to the presence of the exiled French bishops in England. M. Talleyrand further complained that exiled French nobles wore their decorations in public unrebuked. Insomuch as the royal orders of France were no longer in existence, every Frenchman who wore the ribbons of the Holy Ghost and of St. Louis ought to be dismissed the country. Lord Hawkesbury could only reply to this demand by saying that it would be "inconsistent with dignity, honour and the common laws of hospitality" to do anything of the kind. As regards the bishops, the demand appears to have been purely vexatious. As regards the question of decoration, there is this shadow of an excuse, that there are penalties in France for wearing decorations without authority to do so : in England the only penalty is social ostracism. But it is the duty of diplomatic agents to take cognizance of the customs of the country to which they are accredited. In spite therefore of an apparent justification, the French Foreign Office was no more entitled to demand the expulsion of a gentleman who paraded a blue ribbon in public than it was to demand the expulsion of a fugitive monsignore. M. Talleyrand persisted, however, and by the First Consul's orders cited the case of the Pretender, whose removal from France had been accorded at the request of the British Government in days gone by.

In July 1802 M. Otto, the French Envoy in London, complained of the scurrility of the English Press. This is a very natural complaint. Lord Hawkesbury replied, with absolute truth, that in the existing state of English law it was difficult to prove the guilt of an individual so satisfactorily as to obtain the judgment of a court of justice. Unquestionably Lord Hawkesbury was right : and nothing is more certain than that scurrilous statements are least effective when they are ignored.

Nevertheless the French Government maintained the attitude of a highly aggrieved party, and in August 1802 the following demands were made in writing, together with a strong protest against "the deep and continued system of defamation pursued by the British Press, and the odious and degrading terms in which the newspapers were permitted to speak of the French Government." The demands of the French Government were :—

1. The suppression of libels on the French Government published in the English Press.
2. The ejection from Jersey of persons offensive to the French Government.
3. The expulsion from England of the Bishops of Arras and St. Pol de Leon.
4. The expulsion from England of Georges.
5. The expulsion from England of the fugitive Bourbon princes, and their removal to Warsaw ; the English Government to insist on their place of retreat.
6. The expulsion from England of all Frenchmen wearing in public the orders of the ancient Monarchy.

The attitude of the English Cabinet, when face to face with these demands, was moderate. It was pointed out that England was but now at the close of nearly ten years of war. That war had been waged with unexampled bitterness on both sides. England had been compelled to undergo numerous military humiliations, in addition to seeing herself the passive spectator of the horrors of 1793. Vast changes of frontier had been wrought on the Continent, changes which had strengthened France considerably.

On the other hand, the French had to remember the battles of Camperdown, Cape St. Vincent, and the Nile. There was the expedition to Egypt and Syria which had assuredly not resulted favourably for France. In settling a peace after such a war—a war in which ancient states had disappeared or had been broken up, and of which the effects were felt from Washington to Lahore, surely it was only reasonable to overlook such paltry incidents as those which formed the grievance of Count Otto's complaint. So at least reasoned Lord Hawkesbury. While admitting and regretting the appearance of "improper" paragraphs, and even of paragraphs which were "improper and indecent," he could not undertake to commit the law officers of the Crown to a hazardous prosecution every time that the French Government chose to take offence at some wretched scribbler's self-advertising productions. As to sending the French princes to Warsaw, he flatly declined to advise the king to do anything so harsh and inhospitable. He quite admitted that it would be in better taste for exiled notables to cease wearing their decorations in public, but again pointed out that in England it was impossible for Government to take any action in such a matter. The presence of Georges in England he admitted to be undesirable, and steps would be taken to satisfy France in that direction. Count Otto's manner, he commented, was "far from conciliating."

We have now arrived at September 1802, and we are to recollect that in this month Sebastiani started from Paris on his errand to Egypt and Syria, while Sir Alexander Ball on our part and General Vial on the part of the First Consul were engaged in arranging the details of the evacuation of Malta.

In the meantime, however, Switzerland had been occupied and virtually annexed by Ney. Thus before the order of things decreed to be established by the Treaty of Amiens had been brought into existence, a material alteration of the face of Europe was violently effected by the armies of France. It is not possible to accept Bonaparte's view that this action on the part of France was an event of no importance. Under no circumstances can the absorption of one state into another be described as a trifle. As the independence of Switzerland had been guaranteed by the Treaty of Lunéville, it was natural that the English Government should regard the destruction of Swiss independence with profound misgiving. It was regarded in England as another proof that France was careless of the rights of other nations; it was remarked that while France was showing herself punctilious, and even quarrelsome, over small matters at issue, she had no hesitation in raising fresh issues which almost amounted to provocations.

There now reached England a fresh complaint. Stuart, it appears, had told Sebastiani that he could not evacuate Egypt without direct orders to that effect. Lord Hawkesbury hastened to explain (November 30, 1802) that Stuart was mistaken. Egypt had never been annexed, it was at this date an integral part of the Ottoman Empire, and Stuart himself was only remaining in the country until sufficient transport could be collected to enable him to withdraw. Direct orders would, however, be sent to Egypt at once. We may remember that at this epoch Stuart was quiescent in Egypt, merely watching as a disinterested spectator the strife between Turks and Mamelukes. The Anglo-Indian army of occupation had embarked for India five months earlier. So in extreme uneasiness and suspicion on both sides of the Channel the year 1802 came to its close. On January 27, 1803, Lord Whitworth had a long interview with M. de Talleyrand, and the British Ambassador put the whole question of the Press in a nutshell. "Until the First Consul could so master his feelings as to be indifferent to the scurrility of the English prints, as the English Government was to that which daily appeared in the French Press, this state of things was irremediable." This was Lord Whitworth's last word on the subject. The First Consul could never be brought to believe that a government could not control the Press if it had a mind to do so ; the English Government had given the honest assurance of its regret at the vulgarity of the Press, but was decidedly disinclined to enter on a course of hazardous litigation. So this cause of misunderstanding must remain. M. de Talleyrand turned to the question of Malta, and approached it with the utmost seriousness. Lord Whitworth could say nothing, except that the cumbrous provisions of the Treaty of Amiens corresponded to the complicated constitution of the ancient Order of the Knights of St. John, and that the measures leading to the evacuation were proceeding as rapidly as could be expected in the circumstances.

By this time Sebastiani's report had been published, and on February 9, 1803, Lord Hawkesbury wrote to Lord Whitworth on the subject of this " very extraordinary publication," as he described it. This report, so the Foreign Secretary continued, was an official publication addressed to the First Consul and contained the " most unjustifiable insinuations and charges " against king's officers. Moreover, it set forth " views in the highest degree injurious to the interests of His Majesty's dominions." Consequently it would be impossible to enter on any further discussions relative to Malta until satisfactory explanations of the report had been furnished to the British Government.

This despatch summed up and gave expression to the uneasy feelings with which England had watched the conduct of France during the past year. There was no longer a pretence that the Foreign Office did not notice the essentially hostile attitude of the French Government. When it comes to calling for explanations matters have advanced far, and if there be any need for further evidence of the temper of the French Government it would be found in the attitude of M de Talleyrand. On February 17, 1803, Lord Whitworth communicated to M. de Talleyrand the contents of Lord

Hawkesbury's despatch, and was met with the blank denial of any political object connected with Colonel Sebastiani's mission. Sebastiani was merely a commercial agent, so said the French Foreign Secretary. Unfortunately, the colonel's report contradicts M. de Talleyrand in every line. Its tone, as well as the information which it furnishes, is essentially military and political. It is to be observed that at this interview M. de Talleyrand did not deny that the report was an official document officially published; he merely denied the construction placed upon its appearance by the British Government.

From the date of the appearance of Sebastiani's report all hope of preserving peace died away in England. But war was not to come yet; and in the meantime the First Consul sent for Lord Whitworth, and indulged him in one of those famous conversations, in which were mingled menace, frankness and considerable grace of expression. The interview took place at nine o'clock in the evening of February 16, 1803. The British Ambassador was welcomed with sufficient politeness, and was made to hear once more that the two main grievances of the French Government against England were, firstly Malta, and secondly Alexandria. A long monologue ensued, in which Bonaparte made use of the famous phrase that he would rather see the English encamped in the Faubourg St. Antoine than in Malta. He complained bitterly of the scurrility of the London Press and of the safe asylum afforded to Georges. But it was on the question of Egypt that he expressed himself most frankly. Sooner or later, said Bonaparte, Egypt was certain to belong to France. This might occur by conquest or by cession. Whether by force of arms, or by the impending decay of Turkey, Egypt would assuredly fall into the hands of France at no distant date. As regarded England, it was for her to decide on peace or war; if war broke out, the First Consul went on, he was quite determined to invade England, and fully prepared to do so. To this remarkable deliverance Whitworth made no direct or detailed reply; but he did urge, as he had previously urged on Talleyrand, that it was not at this juncture possible to neglect the sensation produced in England by the publication of Sebastiani's report. The First Consul was ready with his rejoinder of mingled ingenuousness and hauteur. Egypt, he said, he could have taken had he chosen ; he might have sent Leclerc to Egypt instead of sending him to San Domingo; his own actions were therefore the best disclaimer that the British Government could desire.

Lord Hawkesbury's reply to the despatch narrating this conversation indicated in detail the difficulties of carrying out the Treaty of Amiens. As regarded Egypt nothing more could be said. The English were only waiting for transport ; and, as a matter of fact, Stuart did actually evacuate Egypt at a moment when the renewal of the war with France was merely a question of weeks. This was an act of good faith to which attention has not been sufficiently directed. As regarded Malta, four nations had already refused to aid in carrying out the provisions of the Treaty of Amiens. These were Portugal, Prussia, Russia and Spain. Until these nations were satisfied that their interests were not materially injured by the provisions of the treaty, no further steps could be taken.

Early in March 1803 a peremptory note was despatched to London. Andréossy, who had succeeded Otto at the French Embassy, stated that he had received express orders to call for explanations on the question of Malta. One would really suppose that no explanation had been furnished. The trick of pretending that nothing has passed when in reality much has passed serves a negotiator often enough, but is a grave discredit to his work when it is examined later. There had, in fact, been given all the explanation that was possible on the subject of Malta; and, quitting the defensive, England demanded, in turn, to know the reason of the steady accumulation of troops in Holland and the north-east of France. M. de Talleyrand, much disturbed in appearance at this movement on the part of England, dwelt on the pacific intentions of the First Consul. But Lord Whitworth brushed these professions aside and told M. de Talleyrand plainly that "security was threatened by the First Consul's views on Egypt." The British occupation of Malta, unavoidable since the Powers could not agree to execute the Treaty of Amiens, would have been continued in any case as a necessary measure of precaution—a defensive operation pure and simple, and wholly indispensable since the First Consul had announced his intention of seizing Egypt. M. de Talleyrand rejoined by submitting a note in which he pointed out that it was "natural" for France to accumulate immense forces around Calais and in Holland, and that such movements could not be taken as in any sense hostile either to England or to Hanover.

In the middle of March 1803 there occurred that famous interview between Lord Whitworth and the First Consul which has been made the subject of controversy, and which has to be examined in some detail in consequence.

Lord Whitworth described it to Lord Hawkesbury in a despatch dated March 14, 1803. On the First Consul approaching the diplomatic circle he indicated the British Ambassador as the representative of people "qui ne respectent pas les traités," reproached him with desiring to force a war on France, and concluded a disagreeable harangue with the words, "ils en seront responsables à toute l'Europe." " He was too much agitated to make it advisable for me to prolong the conversation. I therefore made no answer, and he retired to his apartments repeating the last phrase, 'ils en seront responsables à toute l'Europe.' All this passed loud enough to be overheard by two hundred people who were present, and I am persuaded that there was not a single person who did not feel the extreme impropriety of his conduct and the total want of dignity, as well as of decency, on the occasion." This is an abstract of the dry official narrative. Let us consider the same story as told by one of those who accompanied the First Consul into the room where the diplomatic corps was awaiting him, and who witnessed the scene. The First Consul "walked rapidly into the reception-room and approached the English Ambassador without bowing to any one. He complained bitterly of the action of the British Government. His anger gained on him every moment, and at last reached a point which terrified the circle ; he gave vent to the harshest language and the most furious threats. No one dared to move. Madame Bonaparte and I looked at each other struck dumb with astonishment, while those around us almost trembled. The self-possessed Englishman was put out of countenance, and hardly knew what to say in reply." From these two accounts, written from such different points of view, it is clear that a very unusual scene took place. The obvious explanation of the First Consul's behaviour is that he foresaw the coming out-break and made use of this opportunity for the purpose of throwing the blame on England.

While Lord Whitworth was turning over in his own mind the next morning the significance of the interview, Lord Hawkesbury was signing the despatch which will be viewed as a piece of good or bad faith according to the predilection of the reader, but which was undoubtedly justified by the facts. Writing to Count Andréossy, the French Ambassador at St. James', on March 15, 1803, Lord Hawkesbury pointed out that the Treaty of Amiens was to be understood as referring to the state of Europe existing at the date of the signature of the treaty. That state of things had been materially altered by the action of France. In Switzerland, in Holland and in Italy extensive encroachments had been made upon the territories of inoffensive neighbours. For all this England had no desire to exact compensation ; but Colonel Sebastiani's report it was impossible to overlook. Representations on that head had been "wholly disregarded," and in consequence Count Andréossy was given to understand England could not consent to evacuate Malta.

On March 17 Lord Whitworth reported to Lord Hawkesbury that he could not consent to expose himself a second time to such "disagreeable circumstances," as had attended his last visit to the First Consul. He had, however, seen M. de Talleyrand. The Foreign Minister of the Republic assured him that Bonaparte was inflexible on the question of Malta ;

and Lord Whitworth, in reply, pointed out that the presence of England in Malta was no menace whatever to the power of France, whereas the presence of the French in Egypt (which the British occupation of Malta alone hindered) was a very serious menace to England.

At the end of March Lord Whitworth received the king's command to attend no more of the First Consul's levées, until he had received an assurance that he would not be exposed to a repetition of the treatment which he had received on the last occasion. In the meantime, Count Andréossy had forwarded his explanation of Sebastiani's report. It was, he said, the report of *a* French colonel published in *a* newspaper. If that had been the case there would have been nothing for England to complain of. The point was that it was the report of a special commissioner published in the official organ of the French Government, which was a different matter altogether. It was hardly likely that England would accept Count Andréossy's explanation, but the explanation had been proffered with apparent seriousness, and it was necessary to make a rejoinder. Accordingly on April 4, 1803, Lord Hawkesbury directed Lord Whitworth to state that unless the French Government was prepared to discuss Sebastiani's report, war would be inevitable.\ England was, however, prepared to offer the following terms in the alternative:— Malta to remain in perpetuity in English hands ; Holland and Switzerland to be evacuated by France ; England to recognize the kingdom of Etruria and the republics of Liguria and Italy ; full compensation to be provided for the Knights of Malta and the King of Sardinia.

On laying these terms before M. de Talleyrand on April 7, Lord Whitworth was confronted with the blank repudiation by the Foreign Minister of the official *Moniteur* as a Governmental publication. Face to face with this brazen denial, Lord Whitworth could only report that all negotiations must necessarily be " at a stand."

A week later, however, England had once more to prefer a complaint to the French Government. A violent invective

o

against the British Government in the Hamburg Government Gazette had been inserted by the orders of, and under pressure from, M. de Rheinhardt, the French Minister. Lord Hawkesbury demanded satisfaction for the indignity offered to England in the face of all Europe, and within a week full reparation was accordingly made. In the meantime, Lord Hawkesbury modified the offer of April 4 ; he now instructed Lord Whitworth to say that there could be no further question of specifically carrying out the terms of the Treaty of Amiens : in respect of Malta the conditions had entirely changed. But England would be prepared to evacuate the island after an occupation of ten years' duration, and in consideration of the cession of Lampedusa in full sovereignty. This offer was not unfavourably received at first ; England added the offer to acknowledge the new Italian states, if France would evacuate Holland and Switzerland, and guarantee provision for the King of Sardinia. But on April 23, M. de Talleyrand informed Lord Whitworth that no consideration on earth would induce the First Consul to agree to England holding Malta on any terms ; to the cession of Lampedusa there was no objection. In Lord Whitworth's opinion this was an insufficient concession. "It could never be admitted," he went on, "that the First Consul had a right to act in such a manner as to excite jealousy and create alarm in every state in Europe, and when asked for explanation or security say that it was contrary to his honour and dignity to afford either."

Lord Whitworth saw M. de Talleyrand on the 25th, and again on April 27, but without making way with the negotiations. On April 29, the English Ambassador called again on the Minister of Foreign Affairs, who said that he had not taken Lord Whitworth's instructions to ask for his passports seriously. But Lord Whitworth assured him that "actual war was preferable to the state of suspense in which England, and indeed all Europe, had been kept for so long a space of time."

The interview closed with M. de Talleyrand adding, "J'ai

encore de l'espoir." On May 2, Lord Whitworth asked for his passports; on the 4th, as he was on the point of starting, he received a proposal from M. de Talleyrand to submit the question of Malta to the emperors of Germany and Russia, the kings of Spain and Prussia, and the Batavian Republic for arbitration; or, as an alternative, to surrender Malta into the hands of Russia.

In reply to these advances, Lord Hawkesbury directed Lord Whitworth to point out that they were " loose, indefinite, and unsatisfactory," and that no alternative to the offer of England could be considered. So ended the negotiations between France and England prior to the renewal of the war. It is for the student to decide for himself whether they were inspired by the " treachery, lunacy or weakness " of the British Government, as the Hamburg article declared; or whether that Government bore itself, on the whole, with dignity in difficult circumstances, and with rectitude in the face of questionable claims. The upshot of it all was that war was declared; the Government commanding a majority of 398 to 67 in the House of Commons, and of 142 to 10 in the House of Lords.

It now becomes necessary to follow the course of events in Egypt throughout the year 1803. In January and February 1803, Lower Egypt was strewn with unburied corpses, the results of Osman Bey Bardissi's victorious march. On March 16, the English army was withdrawn from the country and sent to Malta : an act of remarkable good faith, considering the attitude of the French at Constantinople and the menace contained in Sebastiani's report. War between France and England was formally declared on May 22; but as neither Power had any troops left in the country the declaration of war did not immediately affect the destinies of Egypt. In the meantime, if the Mamelukes were divided among themselves the Turks were not less divided. Frequent mutinies broke out, no pay being forthcoming from the capital, and plunder being out of the question. The viceroy lived in retirement at Damietta. Even here he was not safe;

Damietta was stormed and sacked by the Mamelukes in July 1803, and the viceroy sent prisoner to Cairo. Ali Pacha was appointed to succeed him, but could not venture outside Alexandria. In the meantime, the militant religious sect of the Wahábís had actually succeeded in occupying the Holy Places of Islam. This event had the effect of bringing conciliatory offers to the Mameluke Beys from Constantinople. The Holy Places must certainly be recovered. They could only be recovered through Egypt, and the Mamelukes were in no mood to oblige the Padishah. Letters of pardon and compromise were accordingly transmitted to the viceroy. He quitted Alexandria on December 18, 1803, for Cairo, intent on a settlement of the country, based upon the recognition of Osman Bey Bardissi as the lawful ruler of a definite portion of Egypt.

The year 1804 opened with negotiations between Ali Pacha, the viceroy, and Osman Bey Bardissi, the last of the Mamelukes, as he has been called, on this basis. Neither party to the negotiation was altogether sincere ; each charged the other with breach of faith. On January 27, 1804, the viceroy became a prisoner in the hands of the Mamelukes, and was deported to Syria ; he was assassinated on his way thither on January 29, 1804. On February 12 there landed in Egypt Elfi Bey, the rival of Osman Bey, and one of the earliest of those interesting visitors from the East, whom London society has always taken a pleasure in alternately petting and ignoring. Elfi Bey was so far fortunate that he left London before society had grown tired of him, and he landed in Egypt, feeling himself a second Sultan. The favour he had gained in London, and his own undoubted position in Egypt, made him a formidable rival to Osman Bey, who immediately brought about the election of another viceroy to counterbalance the influence of Elfi. Khosrou Muhammad Pacha, a prisoner since the storm of Damietta, was restored to the titular viceroyalty on March 14, 1804. He was deposed the next month, and Khurshid Pacha was set up in his place. The rest of the year passed in similar revolutions, revolutions which in Eastern countries are the invariable prelude either to the rise of a despot or to a foreign invasion : in this case to both. On August 3, 1805, Khurshid Pacha was deposed, and Muhammad Pacha, whose name is better known to us under the Turkish spelling Mehemed, was elected to the viceroyalty.

All this time we are to observe that the Sublime Porte had made only feeble efforts to assert a suzerainty over Egypt. In the course of the six years from 1799 to 1805 three great Turkish armies invading Egypt had been scattered : one by Bonaparte at Aboukir in 1799, the second by Kléber at Heliopolis in 1800, the third by Osman Bey Bardissi at Dámanhúr in 1802. Counting the Syrian campaign of Bonaparte and the losses by sickness, the aggregate loss of men to the Ottoman Empire could hardly have been less than 50,000 men. Even this loss, heavy as it is, was not the measure of damage inflicted on the Sultan by the operations which commenced in 1798. The authority of Constantinople over Cairo had disappeared ; one viceroy had been assassinated, others set up and deposed at the will of the Mameluke chiefs. It was clear that if a really strong man attained to chief power in Egypt, it would be impossible to assert any effective control over him. Mehemet Ali Pacha, much to the astonishment of Europe as well as of Constantinople, proved to be the strong man who was to found the Egypt of to-day. Mehemet Ali was born in the year 1769, like so many of the leaders of the Napoleonic epoch, and was consequently thirty-six years of age when he rose to supreme power. Napoleon, his contemporary, had declared himself emperor in the preceding year. The strength of Mehemet was that he understood both Europeans and Asiatics ; that he treated Europeans as Europeans, and Asiatics as Asiatics ; and that he never confused his methods. The way to deal with Mamelukes was to massacre them ; but this was not the way to deal with Englishmen. Consequently, when General Fraser's army was in the viceroy's power, Mehemet was careful to behave in the most magnanimous and courteous manner. But he did not make the mistake of treating his Mameluke rivals in any such faint-hearted manner. To the end of his long and successful reign there were always two Mehemets in Egypt: Mehemet the modern and enlightened ruler, and Mehemet the ruthless tyrant. He enjoyed the good luck which is a strong contributory cause to so many successful careers. When he assumed the duties of viceroy he was presumed, as we have seen, to be a French nominee; but in point of fact France could do little more than circulate glowing Arabic accounts of the battle of Austerlitz. England supported Elfi Bey, who thought himself King of Egypt, but exercised little direct authority. Osman Bey Bardissi actually held effective control over Upper Egypt. The reign of Mehemet began with three strokes of wonderfully good fortune. The first was the sudden death of Osman Bey, which occurred in December 1806; the second was the sudden death of Elfi Bey in January 1807. The third was, that the English expedition to Egypt, which was designed to place the country under Elfi Bey, did not arrive until March 1807. The failure of that expedition, occurring as it did at the same time as the ridiculous retreat of Duckworth from the Dardanelles, damaged English influence in the Levant. But it also brought home to Englishmen the fact that Egypt was not to be the portion of a spoilt child of London society. Under a ruler vigorous, astute and capable, Egypt was about to become an independent factor in the politics of the Mediterranean. The withdrawal of the English, concerted with Mehemet Ali, strengthened the viceroy's position as the obvious head of the country, and the day after the evacuation of Alexandria Mehemet entered in state the historic seaport, and forthwith summoned the Mamelukes to render to him the homage which had always been refused to the Pachas of Egypt in the past.

His position was not yet strong enough to command obedience; and he did not insist, but spent the years 1808 and 1809 in acquiring vast revenues for himself, in bringing something approaching order into the finances of his country, and in advancing the interests of his family. Mehemet had none of the meanness of soul which rejoices in being surrounded with inferiors. He welcomed capacity ; he trusted it ; he rejoiced to find it in his own family. Far from regard ing his sons with the jealousy said to have been felt by the Effendina of our own day for Tewfik when heir-apparent, Mehemet chose every opportunity of putting his sons into conspicuous and important positions. His plans for consoli- dating his own position and that of his family in Egypt were somewhat disconcerted by the receipt of an order to proceed to Arabia and expel the Wahábís from the Holy Places. If he obeyed this order he would leave Egypt free once more for the re-assertion of the authority of the Sultan, or for a revival of the sway of the Mamelukes. It was possible that he might fail, in which case degradation was a certainty. He therefore attempted to procure his recognition by the Powers of Europe as a merely nominal dependent on the Sultan, and so to gain the position held by the Dey of Algiers or the Emperor of Morocco. But Europe was too distracted to pay heed to Mehemet's ambitions. If he could have secured this advantageous position, it may be safely surmised that he would have returned from Arabia with the Wahábís for his allies, and with his own authority extended over wide provinces which still owed allegiance to the Padishah. If England would have recognized Mehemet Ali he would have declared his independence. In spite of the fiasco of 1807 the British name was greatly respected in Egypt, and the Pacha, taking note of the distractions of the European Powers, would have been contented with the support of England alone. In 1813 he even went so far as to receive with open indifference a present from the Emperor Napoleon, stating in public his preference for the British alliance. As his plans developed, and a large revenue became more and more necessary to him, his respect for England increased. For England, as the one Power capable of blockading either Aboukir, or the Red Sea, could ruin him financially. But it was not to be ; and as his marching orders were repeated somewhat urgently from Con- stantinople he had to consider how he might obey them with the least disadvantage to himself. Of the two dangers—the Turks and the Mamelukes—the Mameluke danger was decidedly the greater. Turkey moved slowly, it might not care to move at all. But the Mamelukes were already in Egypt, united once more in common hatred of a Pacha who was determined to be their master. In March 1811, the viceroy made sure of his position in Egypt by a general massacre of the Mamelukes; and in September, being unable any longer to postpone his departure, he put the Wahábí expedition in hand.

It must not be concluded that English ignorance of Egypt in 1798 implied ignorance of the East. On the contrary, there never was a time when England was so actively concerned in, and so well-informed on, the affairs of Arabia, Persia, Afghanistan and Turkey in Asia. Egypt was the only part of the Turkish Empire about which England had no information, and in which she took no interest. Persia in particular had a threefold interest for England : her intrinsic importance as a great Asiatic Power, her potentiality of becoming a valuable ally in case the Durrani Empire should threaten India, and her important position as the probable heir of the greater part of the Turkish dominions. Even then it was held on all hands that the Turkish Empire could not last much longer. The present shape of the Balkan peninsula, however, was not foreseen ; nor was it anticipated that Persia would be reduced to her present uninfluential position. It was supposed that Persia would wax great as Turkey waned. The idea that both Powers would dwindle under the pressure of Russia had not crossed the minds of contemporary observers. When the Turks were expelled from Europe it was anticipated that they would dissolve into hordes of plundering soldiery, and either become absorbed into Persia or else be stamped out. The greatest danger to India was supposed to be not Russia, but Afghanistan. The doings of French secret emissaries were carefully watched, and the anxiety as to the amount they might have been able to effect in India was repeated outside India when England considered the Wahábís. The influence

of the Wahábís, a warlike, fanatical sect of Islam, was as
menacing in the years 1800–1810, as that of the Mahdi in
recent days. It began in Arabia, but it might extend in any
direction and to any distance.

From all these anxieties there sprang one conclusion. The
British Empire in India must be maintained, and the only
direct menace to that empire was the assault of France. For
Egypt the English had no desire whatever. When it was in
their hands they only asked to be rid of it as soon as possible.
"Every opportunity during my embassy at Constantinople
has been carefully improved towards impressing on the minds
of the Turks how very essentially they are indebted for the
recovery of Egypt to our East Indian interests." So wrote Lord
Elgin to the Secret Committee of the Court of Directors on
February 12, 1802. These views are reflected in the language
of every contemporary observer. Admirals, consuls, military
men, civilians, all held the same views in different words.
Egypt might be, in fact was, the "key to India" in the hands
of a hostile Power. As the "key to India" it could not
possibly be left in the hands of a Power with confessed designs
on India, but all that England asked was that Turkey should
maintain an effective control over the country. And inasmuch
as recent events had shown only too clearly how ineffective
the control of Turkey actually was, we began to negotiate for
the acquisition of Aden. As regards a march on Central
Asia through Syria, it was clear that, if attempted in time of
peace with England, it would be a perfectly feasible operation
for France. There was no Aden in Central Asia which
England could acquire ; and she remained, and must remain,
exposed to that danger. For "Bonaparte's Government will
ever be directed against the Turkish Empire, whether as an
object of immediate conquest, or ultimately in a view to
affect our Indian interests." (Lord Elgin, November 30, 1802.)

Most of these ideas appear to be very wild to-day. When
historians lay it down that Napoleon's European campaigns
were all directed against the British Empire in India, we are
all apt to conclude that these are ingenious theories, suitable

for the lecture-room, but only tenable if we make selections from the facts. On the contrary, the deeper we penetrate into the facts, the stronger and swifter becomes the current of evidence, bearing us on to the conclusion that the history of Europe is inseparable from that of Egypt and India. It is only when we endeavour to treat Europe as if it stood alone that we land ourselves in difficulties.

It was Mehemet's expedition against the Wahábís that made him familiar with the second of the two great ideas which have produced the Egypt of to-day. We must never lose sight of the fact that the Egypt of the eighteenth century was a small province extending no further south than Assouan. Far from Egypt having vague claims, or claims at all, over large tracts outside its own borders, it was, on the contrary, the tribes bordering on Egyptian frontiers who were with difficulty kept at bay. The country was ruled with divided authority; its resources were insignificant. Under Mehemet the divided authority disappeared. There could be no doubt in the minds of European statesmen as to who was the ruler of Egypt. The Padishah could not hope to pit a rival against his viceroy; for the old rivals were destroyed, and the iron hand of Mehemet was on every tribe and every family throughout the country. His vast wealth made it difficult to threaten him and impossible to bribe him. His large family of capable and vigorous sons added strength to his position. What he would do with that position was the problem lying before him when he was compelled to undertake the Wahábí war. On his return from that war his mind was made up: the boundaries of Egypt must be extended by conquest; and they must be extended south. This, then, is the interest, for England, of these obscure years; it is the transformation of a barbarous province into a state formed on a civilized model, with trained armies and fleets, with treasure to fall back upon, a vigorous head and a definite policy not lacking in grandeur.

These being Mehemet's governing ideas—a strong central authority, and a vigorous foreign policy—it may be as well to review the family history which unites the Egypt of Abbas the Second with the Egypt of Mehemet Ali. Three of the great viceroy's sons are known to history—Túsún, Ibrahim and Saïd. Túsún was a remarkable man. It was to him that the conduct of the first Wahábí war was confided in 1812. His character was attractive, and his death in the year 1816 was generally lamented. It was reputed that he died of poison ; more probably he died from excessive indulgence in pleasure after the wearisome campaign in Arabia. Túsún died before he became more than a name for Europe ; Ibrahim was a European figure. The conqueror of Abdallah, the subjugator of the Morea, the victor at Acre, a soldier cool, able and resolute, he did not enjoy the affection of his father, but he was completely in his father's confidence. It is remarkable that Mehemet's reign was disturbed by none of those family dissensions which have marked the rule of most Eastern kings since the revolt of Absalom. Ibrahim the ruthless, Saïd and Túsún alike—very different characters— were unshakably loyal to their father. Ibrahim died in 1839, two years before his father. Mehemet was succeeded as viceroy of Egypt by his grandson Abbas, the son of Túsún. Abbas reigned for thirteen years—from 1841 to 1854—and was succeeded by his uncle Saïd, brother of Túsún and Ibrahim. In 1854, when Saïd became viceroy, Ismaïl was a boy of fourteen, and Saïd was thirty-two years of age. Ismaïl's disastrously magnificent reign did not begin until the year 1863. It lasted for sixteen years, and was closed by his deposition in the year 1879. He died in retirement nineteen years later at Constantinople.

It was through the line of Ibrahim—undoubtedly the most capable of the great viceroy's sons—that the house of Mehemet Ali was to rule over Egypt. Ismaïl was the son of Ibrahim ; Tewfik, who reigned for thirteen years, was the son of Ismaïl ; on Tewfik's death, in the year 1892, his son Abbas the Second, the fourth in direct descent from Mehemet Ali, succeeded to the throne of Egypt. It is true that Abbas the Second does not reign over an Egypt as rich as his great-great-grandfather imagined—an Egypt controlling the whole African coast-line of the Mediterranean, or an Egypt counting Syria and Armenia as tributary provinces; nevertheless, the Egypt of 1900 is four times the size of the Egypt of 1800.

To return to the year 1812. The Wahábís, a religious reforming sect, had developed, by natural process, into a conquering horde. At the time when Mehemet Ali was ordered, sorely against his will, to proceed against them, they held the Arabian littoral both of the Red Sea and Persian Gulf. They were undisciplined; brave as the Mahdists, but not so savage; and they had no great leader. Túsún Pacha effected what was required of him—the recovery of the Holy Places; but the control of the illimitable expanse of the Arabian deserts was clearly beyond his reach. The viceroy himself was not entirely convinced that the conquest of Arabia was impracticable; and he moved from Suez at the end of the year 1813, prepared to find that he could do better than his son. In so far as gaining a great victory went, he certainly did better. But a great victory effected very little in the face of the peninsula in arms. It might be agreeable to Turkey to see a too formidable vassal waste his strength in the Arabian desert, but the viceroy had no intention of losing more men in the service of the Sultan than might be absolutely necessary to carry out the orders of Constantinople. Nevertheless, the situation was embarrassing. To confess himself unable to subjugate Arabia would be to invite the Sultan to depose him. To proceed was ruin. For the moment events in France dictated his actions. Napoleon escaped from Elba; and Mehemet hurried back to Cairo from Arabia in extreme anxiety. He entered his capital on the day after the battle of Waterloo, and began the reform of his army by drilling it on the European model. At first the decree excited too much opposition, and it was wisely withdrawn; but the viceroy's mind was already dwelling on the idea of a re-modelled army as the only possible means by which he could convert Egypt into a great state.

There was no real danger from the side of France in 1815. Only the grandeur of Napoleon's name and the memory of the French campaign of 1798 brought Mehemet back to Cairo in haste. Nevertheless, he continued his preparations for augmenting the army. By the adroit mixture of pressure and persuasion, of which throughout his career he was a perfect master, the viceroy overcame the repugnance of his soldiery to European training. A large, disciplined army would make him independent, perhaps the master of Constantinople; but in 1816 he had not yet a large, disciplined army ; the Sultan's orders for the suppression of the Wahábís had been issued six years earlier, and even Constantinople grows impatient when delays extend to six years.

In this year Ibrahim the Mameluke died. The colleague and rival of Murad the Brave eighteen years before, he had sunk into insignificance, and his death at Dongola created no sensation. Túsún Pacha died too ; and the Arabian command was conferred on the viceroy's son Ibrahim, the great-grand-father of the present Khedive.

The next three years of Egyptian history contain two highly important events : firstly, the suppression of the Wahábís, and secondly, the creation of a disciplined Egyptian fleet and army. The Wahábí war went on slowly but steadily. The arid country, the courage of the enemy, and the vast distances to be traversed made the campaign slow. If the Wahábí leader had been capable the issue might have been different. As it was, Abdallah surrendered to Ibrahim Pacha in the autumn of 1818, and was sent a prisoner to Cairo. The viceroy received him on November 17, and promptly forwarded him to Constantinople, where he was publicly beheaded on November 22, 1818. This dramatic event satisfied the Porte. The Wahábís were not really suppressed, still less was Arabia subjugated. But the execution of Abdallah was undeniable ; for the rest the viceroy was growing too strong to be safely interfered with. Ibrahim was allowed to return to Cairo, and his public entry on December 11, 1819, was made the occasion of according him a triumph. The viceroy was well out of the Arabian imbroglio, and could turn his attention to more profitable undertakings.

An increased revenue was indispensable to Mehemet Ali. Egypt—still the Egypt of 1798, so far as taxable area of territory was concerned—groaned under the burdens imposed by the Arabian campaign, by the withdrawal of so many men from active careers and their absorption into the regiments of a re-modelled army. The creation of a fleet was in itself a ruinous enterprise for a country so sparsely timbered as Lower Egypt. A double relief was therefore effected when the conquest of Dongola was put in hand. This undertaking relieved the finances of Egypt,'and promised an increased revenue. The expedition was entrusted to Ismaïl Pacha, the viceroy's son, who quitted himself well. The battle of Korti, fought on November 4, 1820, tried the mettle of the new soldiery, and proved it to be good. Ismaïl did even better than he was expected to do: to the conquest of Dongola he added the capture of Berber and Shendy and the subjugation of the district around Sennaar. This was a notable extension of Egyptian territory. Unfortunately, while celebrating his victories at Metemmeh on his way home, Ismaïl Pacha was surrounded in his place of feasting by a band of raiders and murdered. Mehemet himself took up the command vacated by the death of his son, and added Kordofan to Dongola and Sennaar, the new provinces of the viceroyalty of Egypt.

It was clear, then, that Egypt could, without difficulty, be turned into a great African Power. But the very ease with which the Egyptian conquests were effected turned the viceroy's mind in another direction. The same year which saw his own remarkable achievements in Upper Egypt saw the revolt of Ali Pacha of Janina. This notable, a man greatly Mehemet's inferior in every respect, nearly succeeded in establishing his independence. He was as well capable of laying up riches as the viceroy of Egypt ; but he was ignorant of the use of wealth, and ruined his own cause by sheer miserliness. He had far more of the cruelty of the Turk—although he was not a Moslem by birth—than Mehemet. If, then, a man so markedly inferior was able to defy Constantinople, what might not be accomplished in Europe by a ruler of sagacity commanding the services of a fleet and a disciplined army ? The revolt of Ali Pacha opened up possibilities for Egypt beside which the conquest of Kordofan sank into insignificance. Nevertheless, Mehemet was not the man to sacrifice the substance to the shadow. He was well aware that interference in European affairs was a different matter altogether from campaigns in the Red Sea or on the Upper Nile. He could hardly appear in Europe as a conqueror; and meantime, Upper Egypt and the Soudan lay open to his arms. He pushed his armies as far south as Khartum, which was occupied in September 1824, and would doubtless have proceeded further but for the rapid development of political affairs in the Grecian peninsula.

In England we are accustomed to say that the sympathy of England with the cause of Greek independence was based, partly upon a generous enthusiasm for the cause of liberty, and partly upon a love of Greece derived from our classical studies in boyhood. On the Continent it is more usual to say that it was gratifying to England to see Turkey and Egypt weaken themselves in vain endeavours to subjugate Greece while English merchants made large fortunes by supplying both sides with arms and stores. France was the patron of Mehemet Ali, and one of the signatories of the Treaty of London to be noticed shortly, and would be advantageously placed in any event. Russia was much influenced by the counsels of John Capodistrias, but was, on the other hand, deeply pledged to the principles of the Holy Alliance. Here we have, in brief, all the influences which were at work when, on March 6, 1821, Ypsilanti crossed the Pruth, entered Moldavia and captured Jassy in the name of the Greek Revolution. He counted upon the support of Russia. But secret societies are not agreeable to Tsars, and Ypsilanti was left to his fate. He died in prison at Theresienstadt shortly afterwards. Nevertheless, his audacious move had set the Greek Revolution on foot. The prompt massacre of Greek Christians at Constantinople did not avail to check its progress, and during the years 1821, 1822 and 1823, while Egypt was advancing on Khartúm, the Morea was in insurrection from end to end. It was in order to suppress this formidable revolt that the Sultan decided, towards the end of the third year of its duration, to call in the assistance of his powerful, his far too powerful, vassal. On January 16, 1824, the suppression of the insurrection was entrusted to the viceroy ,of Egypt. Mehemet Ali was created Pacha of the Morea.

If Mehemet had somewhat reluctantly obeyed his sovereign's order to attack the Wahábís, it was with far different feelings that he received the order to subjugate the Morea. The one order took him far away into unknown deserts, and pledged him to a campaign from which he was fortunate to have escaped with credit. The new order brought him publicity and made him a European figure. Of the directions in which the boundaries of Egypt might be extended, Europe was by far the most attractive to Mehemet. Let us once more review the possibilities of conquest. They might be Asiatic, either in Arabia or Syria. After the Wahábí war, Arabia was clearly out of the question ; Syria was Mehemet's latest enterprise. They might be African, either by extending the boundaries of Egypt south (which was possible and profitable, but led the viceroy still farther away from Europe), or by obtaining recognition as a Barbary State, or by assuming the suzerainty over the other Barbary States, as we saw in his correspondence with Prince Polignac. Finally, they might be European. In this direction lay the greatest possibilities, but also the greatest difficulty : how to get to Europe? The command in the Morea solved this difficulty, which, without the command, amounted to an impossibility. Encamped on European soil, possessing a fleet already nearly strong enough to face the Turkish fleet with success, commanding a strong body of disciplined troops, of whom some had already seen active service, with nothing stronger to oppose him than the Janissaries (not yet massacred), Mehemet was justified in believing that only the most untoward events could long delay him transferring his capital from Cairo to Constantinople. The

most untoward event it actually was which befell ; but nearly
three years were to pass before the battle of Navarino threw
Mehemet back upon Egypt.

Mehemet Ali, Viceroy of Egypt, Pacha of the Morea,
entrusted the conduct of the campaign in Greece to Ibrahim
Pacha, his son. For some time past the viceroy had borne
himself more as a sovereign than as a subject ; and he now
deputed his functions as generalissimo by land and sea in
truly royal fashion. In so far as the great Powers were concerned England had long been indifferent to the personality
of the Egyptian ruler. So long as he was visibly independent of France he might wax as powerful as he was able.
France, however, had sedulously cultivated friendship with
Mehemet. Even at this date, ten years after the fall of
Napoleon had deprived France of all chance of directly
dominating the affairs of Egypt, that "indirect influence" which
was to culminate, forty years later, in the visit of Napoleon
the Third, had made considerable progress. Mehemet's
sympathies were pronouncedly French. At the eastern end
of the Mediterranean the interests of France were sure of
attentive consideration in the hands of the viceroy. At
the western end of the Mediterranean, France was already
asserting herself in Algiers. In the centre of the Mediterranean, death had removed the strong hand of Maitland, the
ruler of Corfu and Malta. France was in the ascendant once
more.

That development of the arts of destruction and defence
which ensued with the opening of the age of inventions
had not at this date materially altered the conditions of
naval warfare. It is out of the question to-day for any but the
most wealthy nations to maintain powerful fleets. Greece, like
all other small countries, must rest contented with a naval
inferiority so marked as to be absolute. It was not so in
1824. The Greek insurgents counted many brave sailors in
their ranks, and easily equipped a fleet capable of showing
fight to the Egyptians. Ibrahim Pacha was compelled to
winter in Candia. It was not until March 1825, that he could

P

clear the way for his transports and land troops in the Morea. The better discipline of the Egyptians told at once, even on ground strange to them and favourable to the Greeks. Navarino was captured in May 1825, and Ibrahim proceeded forthwith to lay siege to Missolonghi.

Missolonghi is inseparably associated with the name of Byron, who had died there in the previous year. He was one of the most distinguished of many Englishmen who sympathized actively with the Greeks. We must not forget in reviewing the career of Mehemet Ali, the excitement aroused in Europe by the sudden appearance of a new, a formidable, an intelligent, a belligerent Power invading Europe successfully in the name of the faith of Islam. The Morean expedition evoked the opposition of all who cared for Christianity, of all who admired the struggle of the weak against the strong, of all who cared for culture. Every argument that could appeal to an Englishman told heavily in favour of the Greeks and against Mehemet. Time, therefore, was on the side of the Greeks. Their own prowess and the deliberate movements of the Turks prevented Ibrahim from making that rapid progress which could alone have ensured his success. Missolonghi did not fall till April 1826, by which time great events were in progress elsewhere.

Of the three Powers, Russia, France and England, only one could move. England, burdened with the enormous National Debt of eight hundred millions, anxious about the succession to the throne, in financial distress, and distracted by the cry for Reform, bordering on an internal revolution, was less fit than ever to embark on an adventurous policy in the Mediterranean. Beyond "moral support" and private subscriptions, Greece could expect nothing from England. In France the restored Bourbons were not in great favour with the people; and so far as the Mediterranean was concerned, Algiers was, as we have seen, the all-important question. There remained Russia. A new Tsar, the Emperor Nicolas the First, had recently ascended the throne. In the emperor's strong character two features were prominent—a profound belief in the righteousness of absolute government, and a genuine religious fervour amounting to fanaticism. As regarded the question of Greek independence, there could be no doubt on which side his influence would be exerted if he decided to interfere—that is, if he consulted his religious feelings only. But, on the other hand, the Sultan was an absolute sovereign, and the Greeks were in rebellion—and rebellion against the authority of an absolute sovereign ought not to be encouraged. The traditional policy of Russia, the religious duty of warring against Islam, turned the scale. The Greeks must be supported. Young, powerful, with a profound belief in himself, the Tsar Nicolas came to London and set on foot those negotiations which were to result in the rise of a new European state—the first of the five which have now been carved out of Turkey in Europe.

In the meantime the Sultan had paid his great vassal the compliment of imitating his policy. The Janissaries were to the Sultan what the Mamelukes had been to the Khedive. Mehemet had begun the making of modern Egypt by massacring the Mamelukes. In June 1826 Sultan Mahmúd massacred the Janissaries. The strong position of Egypt was in great measure due to the possession of a disciplined army; with a disciplined army Turkey ought soon to be still more powerful. So reasoned the Sultan; but he overlooked several considerations. Mehemet had harried the Mamelukes for some years before he ventured to attack them. For the massacre he chose the moment when the attention of Europe was fully occupied elsewhere, and when nobody cared whether the Mamelukes existed as a political body or not. The massacre of the Janissaries took place when the Sultan's dominions were aflame with rebellion and honeycombed with discontent, and when the hereditary foe of Turkey was clearly meditating a move. Consequently, though the measure may be applauded as an act of courage, if of ruthless courage, it was hardly politic. The massacre, moreover, was useless unless followed up by the creation of a new army; and the work of reconstruction proceeded so slowly that the Russian war, soon to be declared, found Turkey practically without any army at all.

On July 6 1827, just thirteen months after the massacre, the Treaty of London was signed by Lord Dudley, Prince de Polignac and Prince Lieven. It consisted of seven articles, and provided for mediation by the three contracting Powers. Greece was to acknowledge Turkey as suzerain, and pay a tribute, but to be otherwise uncontrolled from Constantinople. Other details were to be settled later, but instructions to this effect were to be forwarded forthwith to the Mediterranean. The contracting Powers were to abstain from any acquisition of territory at the expense of Turkey; all arrangements decided on by the Powers were to be guaranteed by them. There were three additional secret articles. If the Porte declined to accept the mediation of the Powers, Greece was to be recognized by the appointment of consular agents to represent the Powers, who would further "use all the means which circumstances might suggest" to bring about peace without actually taking part in hostilities. The Powers would moreover continue with their arrangements for re-organizing the Greek state, irrespective of the attitude of the Porte. A Joint Note calling upon Turkey to assent to these conditions was presented on August 16, 1827, and rejected point blank. The Reis Effendi, with much appositeness, quoted the motto "Dieu et mon droit" and asked how the English, whose motto it was, could present a demand so preposterous. Orders were issued to the three Admirals commanding in the Mediterranean to hinder the landing of troops and stores, but on no account to make an attack upon the Turkish or Egyptian ships of war or transports.

Here we have a situation in which serious mischief is almost unavoidable. The Admirals were forbidden to open fire, but were enjoined to arrest the military operations of the Egyptians. A commanding officer holding contradictory instructions of this nature will probably be guided by his own sympathies, trusting to be excused for possible indiscretions on the ground that he has acted "in the spirit of his instructions."

This is what occurred at Navarino. The blockade of Navarino was partly effective. It closed the harbour to the Egyptian fleet, and deprived Ibrahim of the immense advantage conferred by the almost unchallenged command of the sea. It could not, however, prevent Ibrahim pursuing his operations on the mainland around Navarino with much success. Clearly, if the sympathies of the Admirals had not been enlisted on one side, the obvious and only course for them to pursue was to report on their embarrassments, to request further instructions, and to point out that their actual instructions were nugatory without permission to fire on the Turkish fleet. The alternative would have been to land a composite force for joint military operations on the mainland. The Admirals did not take this course. For a whole month—September 25 to October 20, 1827—they continued making remonstrances, standing in and off, and observing the progress of Ibrahim's land operations. On October 20 they fought the battle of Navarino. The Turkish and Egyptian fleets were destroyed, and the independence of Greece was assured. But the action was fought contrary to instructions and in violation of the terms of the Treaty of London. The historic description of the battle as "an untoward event" appears, in the light of history, to have erred on the side of moderation. At the time the news of the battle was received with enthusiasm, an enthusiasm not perhaps wholly justified by the course of subsequent events. Perhaps the best parallel to the state of things on the Greek coast in 1827 will be found in the state of things in Cuba seventy years later. It was true that the Sultan was in his rights in suppressing the revolt in Greece : just as Spain was within her rights in taking action in Cuba. But so great a measure of weakness in warlike operations amounts to serious inconvenience to other Powers—as the United States discovered. In 1827 Turkey had been for six years endeavouring to subjugate a people who, even to-day, number only two million souls. For the last three years Egypt had been drawn into the struggle, and there seemed no chance that the end of the struggle was approaching. All these are palliations; but the fact remains that the attack was unauthorized. Consequently, it is hard to refuse sympathy to the Turkish demands presented on November 10, with commendable promptitude. These included (1) the immediate withdrawal of the allied forces from interference in the war, (2) compensation for the loss inflicted on Turkey by the destruction of the Sultan's fleet, and (3) ample satisfaction to the Porte for the insult of breaking the peace with a friendly Power. These demands were all refused, and the Ambassadors of France, Russia and England quitted Constantinople on December 8, 1827.

We are less concerned with the morality of the battle of Navarino than with its political effects. These were firstly to lay Turkey open to the assault of Russia, which took place in the following year. The war of 1828–29 was the first of the three attempts on the part of Russia to break through to the Mediterranean. On this occasion it was clear that France and England had played into the hands of Russia. Without the help of Codrington and de Rigny, the Russians could not have attacked, still less destroyed, the combined fleets of Turkey and Egypt. France and England, then, profited nothing by the battle of Navarino; so the next attempt of Russia to break through to the Mediterranean found France and England united against Russia; and the result was costly and damaging. The third attempt made after the usual interval of peace was still more disastrous; for it definitely closed the Balkan peninsula to the advance of Russia. The next attempt, according to appearances, will be made with France on the side of Russia.

The second result of the battle of Navarino was considerable damage to the good name of England. Judged by the canons of conduct by which the English professed to be guided, their action was a treacherous violation of good faith towards Turkey, and an inexplicable breach of the peace. As premier naval Power England had taken a leading part in the massacre, which Turkish critics can hardly be expected to distinguish from the massacre of Sinope. Consequently, though French and English men-of-war had alike fired on the Turkish flag in time of peace, English reputation had suffered most damage.

The third and most important result of the battle of Navarino was that Mehemet was nearly ruined. Further, it was clear that he could look forward to no more enterprises on the coast of Greece. He could not afford to equip another fleet. His enterprises must, therefore, be land enterprises. When the Russian war broke out the Sultan called on the viceroy to support him. But the viceroy urged, plausibly enough, that the three years' campaign in Greece had greatly strained his resources, and that the battle of Navarino had left him helpless. Egypt could do no more. Consequently the Sultan was left to struggle as best he might with Diebitsch, whose successful campaign terminated in November 1829 with the Peace of Adrianople.

During all this time the viceroy had pushed on his armaments with all the speed possible. There is little doubt that he could have afforded substantial assistance to the Sultan if he had been so minded. But a weak Turkey suited his ambitions very well. In the campaign of 1824–27 in the Morea his object had been, not to strengthen the Sultan, but, having gained a footing on the mainland of Europe, to supplant him. He would have succeeded but for Navarino. His object in 1828–29 was to recover from his losses, and to hold himself ready for the next opportunity. This occurred in the embarrassments of France in dealing with Algiers.

As we have seen, the viceroy, counting on his favourable standing with the Court of France, proposed to take over the Barbary States, provided that he was subsidized to the amount of twenty millions of francs, and was supplied with ships to guard and convey his troops. But if Mehemet had designs on Islam, France had designs also. To control the African littoral was a grand design ; but it was the design of France as well as Egypt. It is true that Charles the Tenth was prepared to share the littoral with Mehemet ; and while himself occupying Algiers, was ready to leave Tunis and Tripoli to Egypt.

But this would have made Mehemet's boundaries conterminous with those of a great European Power. He knew too well what that state of things implied ; he rejected the compromise, and entered on that course of action which was to make the " Eastern Question " a controlling factor in the politics of Europe. A controlling factor it remained, until the Eastern Question itself merged in the greater outlines of an imperial policy.

Arabia, the Soudan, the Morea, the Barbary States—these were the four directions in which Mehemet had endeavoured to extend the boundaries of Egypt. Arabia had no potentialities ; the Soudan had potentialities innumerable, but it led him far from the heart of Islam ; the Morea was an ideal opening, but he had been thrust back by Christendom united ; France had headed him off from the Barbary States, and Mehemet was now sixty-one years of age. If his ambitions were to be realized in his own lifetime, there remained but one opening. In the month of November 1831, he moved on Syria, resolved to approach Constantinople by land, where there could be no danger of a second Navarino.

PART III

THE STRUGGLE FOR ITALY

III

ENGLAND was directly concerned in the affairs of Naples and Sicily throughout a period of twenty-two years from 1793 to 1815. These intimate relations sprang from the horror with which both kingdoms regarded the execution of Louis XVI. of France. The sister of Marie Antoinette, Maria Caroline of Naples, concluded, on July 12, 1793, that alliance with England which she was to find alternately the source of so much exultation and such bitter disappointment; which was to be adorned by the romantic career of Nelson, and to be darkened by the tyranny of Lord William Bentinck.

It may be well to consider at the threshold of this study what was the precise importance of the kingdom of Naples and Sicily in the affairs of Europe. Naples (as it was often called) was a considerable European state of the second class. It was by far the most important of the states of Italy, and in so far as the affairs of the Mediterranean are concerned, the island of Sicily, which formed the southern division of the actual kingdom, was of greater importance than the mainland province for obvious reasons of maritime strategy. The kingdom, to be attacked with any effect must be attacked by sea; and even on the northern land frontier it could not be approached except through states whose subservience or, at the least, neutrality must be secured. Malta might with fore-thought be made a *point d'appui* for the attack of Sicily; but it is far more to the point to say that Sicily unfriendly neutra-lized any strength conferred by the possession of Malta. It is clear that the kingdom of Naples and Sicily, although of secondary importance as a European state, was of great importance in all matters of policy connected with the Mediterranean.

The offensive alliance with England did not at once plunge Naples into war with France. More than a year passed before, on October 18, 1794, war was declared. The declaration coincided with, and was perhaps encouraged by, the British occupation of Corsica, which was at this date emphasized by the formal proclamation of George the Third as King of Corsica, and the appointment of Sir Gilbert Elliot to be his viceroy resident in the island. Here, then, was a military position serious enough for France : for the hostile frontier of Naples was advanced, and her naval power materially increased by the presence of the British fleet. On the side of the French, however, there was the military genius of Bonaparte to be reckoned with; while the allies had nothing to oppose to the French except the incompetence of Hotham, the disorganization of the Neapolitan armies, and the disturbed condition of Corsica brought about by the treason of Pasquale de Paoli. There could be but one result in the face of odds so unevenly balanced ; the defeat—though not altogether inglorious—of the Neapolitan army. An armistice was arranged on June 5, 1796. The armistice was followed, three months later, by a peace, the terms of which require some attention.

France was not, at this date, in a position of overwhelming strength face to face with Naples. We are removed by ten years of eventful history from the date when Napoleon could depose the Bourbons of Naples with a stroke of his pen. In 1796 France must needs be contented to treat with Naples on something approaching equal terms. An advantage had certainly been won by the war, but not an overwhelming advantage, and it is instructive to observe in what direction the advantage was pressed. No cession of territory was demanded, but the British alliance must be dissolved, and all Neapolitan ports closed to the enemies of France. Thus the offensive power of Naples, considerable only when maritime operations were projected, was destroyed. More than that, the offensive power of England was seriously prejudiced. Her position in Corsica became untenable with the whole island in revolt, and only fourteen sail of the line to confront three times that number of hostile ships available for warfare in the Mediterranean.

The advantages gained by France by the peace with Naples were thus much greater than appeared on the surface. The terms of peace were so little oppressive that there was a fair chance of turning a foe into an ally. Henceforth and for a time, the neutrality of Maria Caroline was respectful. That sufficed for Bonaparte, in whose opinion Naples was a power to be dealt with later and at leisure. For the moment his principal object was to secure the withdrawal from Mediterranean waters of the British fleet. In confusion and alarm the English evacuated Corsica and concentrated on Gibraltar. The way was now clear for the Egyptian expedition, and the favourable Treaty of Campo Formio, concluded in October 1797, settled Italian affairs for the moment. Venice disappeared as an independent state, Bonaparte taking care to secure for France the Ionian Islands, an acquisition which aroused but little suspicion. In so far as the transaction was commented on, it was chiefly with surprise at France being pleased to relinquish the rich mainland possessions of Venice into the hands of Austria, and to rest contented with so insignificant a share of the spoil. Naples made a half-hearted attempt to assert some interest in their reversion, but the pride of the Neapolitan Court was, not unnaturally, gratified at the consideration shown to her representations in other directions, and the Powers watched without misgiving the incorporation of the Seven Islands as a department of the French Republic.

In these negotiations Bonaparte alone pursued an intelligible object with a single eye. To him the Ionian Islands were priceless. In the hands of Venice they had possessed but little value, but in the hands of France they were immediately transformed into an arsenal and a recruiting ground. They became a formidable outpost of the Republic, and a starting-point for Bonaparte's undisclosed and unsuspected operations for the conquest of the East.

The year 1798 is filled with the record of that momentous campaign, but even while the campaign was in progress the policy of France towards Naples developed in a remarkable manner. At first, and while the Egyptian expedition was undisclosed, and the undisputed command of the sea was indispensable to the French, Naples was treated with distinction. It is whimsical to observe that the good feeling of the queen was won by surrendering into her hands the two principalities of Benevento (from which Talleyrand was afterwards to take his title), and Ponte Corvo (which the future Prince of Ponte Corvo was only to relinquish for the crown of Sweden). As the year wore on, and the necessity for keeping on good terms with Naples became less urgent, the tone of France changed considerably. On July 24, 1798, the Ambassador of the Republic demanded the exclusion of the English from all Neapolitan ports, and the surrender of the harbour of Messina into French hands. The real policy of France had not been unsuspected in Naples ; for two months before these demands an offensive alliance with the Court of Vienna had been brought about through the instrumentality of Thugut.

Whether this secret action of Naples had affected France or not, the demands to be faced by the Neapolitan Court were plain and peremptory. Would the queen yield to France the command of the Mediterranean? Would she permit Bonaparte to knit Toulon to Corfu by Messina (supporting his latest acquisition of Malta), and so encircle her kingdom with a chain of strong maritime posts, or not? The queen had no hesitation. It was not for this that she had presented Bonaparte with a gold snuff-box. The principalities of Benevento and Ponte Corvo would be dearly bought indeed if their price was to be the abdication of the throne of Naples and Sicily. Moreover, France was no longer the France of La Touche-Tréville. The battle of the Nile was a grave

factor in the situation, and when the Queen of Naples, far
from acceding to the demands of Garat, welcomed and suc-
coured the British fleet in her harbours, the net result of the
years of negotiations and operations proved to be this—that
the land-power of Naples had been strengthened, her influence very considerably increased, and the bonds of alliance
with England drawn closer as the natural result of the
ineffectual menaces of France.

Garat was recalled and La Combe St. Michel presented his
credentials on October 2, 1798. La Combe St. Michel was
a regicide, and his appointment had the effect of stimulating
the activity of the Republican clubs of Naples. It was also
one of the most disagreeable appointments that could possibly
have been made in so far as the queen personally was concerned. It was not likely that menace and insult, especially
when delivered by a man of the stamp of La Combe St.
Michel, would outweigh the charm and the glamour of the
English alliance as presented by Nelson and Lady Hamilton.
If Garat was unsuccessful, La Combe St. Michel was still
more unsuccessful. The English alliance was formally proclaimed, and a month later Sir Charles Stuart captured
Minorca. France, all powerful in the Mediterranean in 1797,
was fast falling back to a secondary position.

The policy of Naples, directed by the energy of the queen,
and served by the genius of Acton, was at this epoch to take
the offensive whenever possible. The queen was under no
illusion ; she was fully aware that France, with Bonaparte to
command the Republican armies, was a terrible foe to provoke.
But she decided that since the design of France was nothing
less than the extinction of Naples, it was giving her kingdom
the best chance in the unequal struggle to deprive France of
the advantage of choosing the most convenient moment to
declare war. There was, at Naples, none of the besotted self-
complacency which pervaded the Court of Prussia. No one
could have ventured to say in the queen's presence that " His
Majesty's army could show several generals who were the
equal of M. de Bonaparte." It was wise of Maria Caroline

to pass over the aspirants to the chief command in her own army, and it was nothing more than the deplorable bad fortune that pursued her through life that sent her Marshal Mack from Vienna. England made the cause of Naples her own; Russia was a staunch ally, if an ally whose arms were somewhat difficult to render effective; and at the end of November 1798 the troops of Maria Caroline crossed the northern boundary of her kingdom and occupied Rome. In this brief and disastrous campaign Mack and King Ferdinand were opposed to Championnet and Joubert. Within three weeks the Neapolitan troops were expelled from Rome and driven in headlong rout across the frontier. At this date, for the first time, the essential weakness of Naples stood revealed. In the summer of 1798, to the casual observer, Queen Maria Caroline appeared a powerful monarch. But the events of the autumn unchained the long-suppressed passions of the lazzaroni. Their excesses stimulated and gave apparent justification to the activity of the Republican clubs. In a fortnight Naples was in an uncontrollable ferment, in which the royal authority disappeared altogether. It was no longer a question of invasion or resistance to the invader, but of preserving life if possible. That service Nelson was able to render to the distracted queen, and on the last day of the year 1798 the sovereigns of Naples and Sicily embarked on H.M.S. *Vanguard* and fled from their capital. Naples was lost; there remained Sicily.

The year 1799 opened favourably for France. What was lost at sea appeared to have been regained on land. The Egyptian expedition had failed, it was true. The Ionian Islands were menaced, Minorca had fallen into the hands of the English, and Malta was blockaded. But, on the other hand, the most formidable Italian opponent of France was in exile, and the territory of Naples lay open to the advance of the Republican armies down to the Straits of Messina. Moreover, what had happened on the mainland might happen in Sicily. If Palermo followed the example of Naples there would be much to compensate France for the battle of the Nile.

The history of the next sixteen years was foreshadowed in outline by the events of 1799. It was evident that France could advance through the mainland unresisted, except in so far as the invasion might arouse the national feeling of the Neapolitans. Russia and England might, perhaps, render in the future more help than they had offered in the campaign of 1799, but substantial support was, on the whole, not to be expected. On the other hand, France, recovering with wonderful rapidity from the battle of the Nile, was far superior to Naples at sea. But under the treaty of December 1, 1798, the British fleet was at the service of the queen. Naples, therefore, with some reserves, remained henceforth at the disposal of France; but Sicily was beyond her reach, unless the British fleet were subdued or withdrawn from the alliance. Further consequences flowed from this conclusion. Without the occupation of Sicily it was clear that France must lose Corfu and Malta. It was also obvious that a French dominion, even when established on the mainland, would find itself encircled with a ring of hostile posts; and that the occupation of Naples, even if peaceable, could play no considerable part in the Eastern campaign.

The genius who had conceived that campaign was, in early 1799, entangled in the sands of Syria, while his lieutenant, Desaix, was being lured up the Nile by Murad Bey. Lesser men conducted the operations in Italy. Championnet entered Naples on January 23, 1799, and nominated Rey to be governor of the city. The conquest had cost him, perhaps, 1000 in killed and wounded. The Neapolitan army was scattered to the four winds, but an irregular warfare, a pale anticipation of the uprisings in Spain ten years later, made progress difficult and sometimes costly. There was much that was worthy in the national sentiment of Neapolitans. Among the better sort there was a genuine and intelligible impatience at the tyranny of the Court. The Government had been good in the sense that it was strong and orderly, but its rigid denial of anything approaching freedom of thought, speech or action left the nobler part of the nation disaffected. There is no doubt that the French were welcomed with sincerity by large numbers of the better classes. The nobles appear to have been hardly worthy of their position. Patriotism, the determination to expel the invader and settle national affairs by national measures, was lacking throughout the whole community. The passions of the lazzaroni alone represented national sentiment; and the lazzaroni would as yet yield to no discipline and obey no leader. The form of government now established by Championnet displeased everybody. "The Parthenopœan Republic" angered the populace, not because it was a republic, but because it was pagan in name, although they were themselves pagan in all but the name. It was distasteful to the nobility, not because it was pagan, but because it was a republic, where nobles could find no place. The residue, satisfied with any government that was neither Bourbon nor royal, was not influential or numerous enough to give it stability. From the streets and the country-side there arose a murmur with a presage of danger in it. The discontented called themselves the "Sanfedisti," a cry that came to mean much later on.

In this chaos the idle drafting of a Parthenopœan constitution, the ineptitudes of Faypoult, the supersession of Championnet himself, were incidents that altered nothing. Although a new French commander-in-chief appeared—Macdonald, afterwards Duke of Taranto—the situation remained unaltered on the mainland. But by now the Russian alliance (signed in St. Petersburg on November 29, 1798, the day on which King Ferdinand entered Rome) had become effective. Strengthened by the adhesion of its traditional enemy, Turkey, the fleet of Russia was rapidly reducing the Ionian Islands. Corfu itself yielded on March 1, 1799, and the oddly-assorted allies — Russians, Turks, English and Sicilians—were now in a position to attempt a descent on the mainland. A man of genius forestalled them—Cardinal Ruffo —who landed alone in Calabria in February 1799, and using the authority of the king summoned round him an army to whom he gave the welcome rallying-cry, "the Holy Faith."

It was at this date that Bonaparte started from El Arish for the conquest of Syria. Against Ruffo, churchman, soldier, noble and Neapolitan, the French armies could effect nothing. His army of irregulars was stiffened by a detachment of Russians landed on the Adriatic coast two months after his own appearance in Calabria. By the end of April, Gaëta and Naples alone remained to the French, and even in the Bay of Naples a British squadron menacingly cast anchor.

It must be admitted that the least promising ally of King Ferdinand was the ally who rendered him, at this juncture, the most effectual aid. An alliance between Russia and Sicily appears at first sight to be hardly practicable. Nevertheless it was Russia who in 1799 played everywhere the leading part in Italian affairs. Uschakoff's marines landed on the Adriatic coast and enabled Ruffo to clear the countryside. Suvarov simultaneously conducted his famous campaign in Northern Italy. In May 1799 the French were compelled to evacuate Naples after an occupation lasting three months and a half.

The Government left behind after Macdonald's departure was Republican, childishly incapable, grotesque, sentimental and living in a world of dreams. It was naturally opposed to Ruffo and his Sanfedisti, who had from the outset acted in the name of King Ferdinand. It prepared to give battle to the Cardinal, and was supported by the garrison which Macdonald had left behind him in the castle of San Elmo. After a short resistance, the Republicans were overpowered by the Cardinal's Turks, Russians and Sanfedisti, and Naples was re-conquered for King Ferdinand by June 15, 1799. On this day Bonaparte made his "triumphal" entry into Cairo after what he was pleased to call the conquest of Syria. The French did not evacuate San Elmo until July 5. Capua and Gaëta followed the example of the capital and capitulated to King Ferdinand immediately afterwards. It is hardly possible to exaggerate the atrocious incidents of that one fortnight of internecine strife in the streets of Naples, and although such excesses naturally aroused a thirst for revenge, it is painful to remember that Cardinal Ruffo's commission was cancelled. Not all his successes could atone for his having interceded for the "insurgents." The senseless reprisals of King Ferdinand lasted for nearly a year, and brought thousands of his Neapolitan subjects—some of them demonstrably innocent—to the scaffold. To this period belongs the execution of Admiral Caracciolo.

In the meantime the war in Northern Italy was raging. The victories gained there, once of world-wide fame, now fallen somewhat into the back-waters of history, recall the great names of the Archduke Charles, Kray, Suvarov, Massena, Bellegarde, Joubert, Moreau. As one result of these manœuvres, the French evacuated Rome on September 29, 1799, after a six months' occupation, while Bonaparte was anchored at Ajaccio on his way home from Egypt. The year 1799, which opened so hopefully for France, closed with the disappearance of French armies from Italian soil—Genoa alone remaining in their hands. For this result sufficient justice has not been done to the genius of Cardinal Ruffo. It is true that his victorious operations were only possible because Russia, Turkey and England secured for him the command of the sea. But without a leader whose name and presence conciliated the national and religious prejudices of the Neapolitans, Russian, Turkish and English troops would have been able to effect but little. To the Neapolitans these strange allies were as much aliens as the French; the Russians and English were equally heretic, and the Turks were infidels. The Cardinal, with his rallying-cry, "La Santa Fede," had effected wonders in this campaign—a campaign at once brilliant, grotesque and futile: brilliant for its astonishing success, more grotesque for the composition of its armies than any of the Crusades, and entirely futile, because on November 9 General Bonaparte was elected First Consul, and proceeded to take over the conduct of the French operations in Italy. The Eastern expedition had failed. Bonaparte had been unable to hold the line of the Nile, he had been compelled to retreat from Syria. It was useless to have slaughtered thousands of Turks at Aboukir when millions remained and his own army dwindled daily. His daring bid for the sympathy of Islam had failed; he had been unable to recruit native levies. These difficulties were not, as he himself pointed out (erroneously perhaps), greater than those which the English had surmounted in India. But one defect was irremediable— he had lost his base. When he started he could calculate on the command of the sea and the possession of Corfu; Malta he took in his stride; the mainland of Italy displayed a friendly neutrality. All this had changed in the course of eighteen months' warfare. It was necessary to recommence operations *ab initio,* and to undertake the re-conquest of Italy, whether for its intrinsic value, or for revenge, or from personal ambition, or as an indispensable step in any future operations in the East.

As regards those future operations, we have accustomed ourselves to consider them as wild imaginings, to comment on Bonaparte's ignorance of naval affairs, and to date the utter ruin of his schemes of Eastern empire from the battle of the Nile. Contemporary observers were of a different opinion. To Lord Elgin, the newly-appointed Ambassador at Constantinople, the situation appeared far from reassuring. He brought from England the latest public opinion of the country, which was cast on the gloomiest lines, and he had access to the best sources of information open to public men in the East. So far from considering that the battle of the Nile had closed the door of the East to the French, he wrote to Lord Mornington, the Governor-General of India, that he was instructed (and considered it as his first duty) to keep his eye on the East. It was under special orders from Dundas that the British Ambassador at Constantinople opened up the correspondence with Lord Mornington. This is hardly to be wondered at, seeing that Tipu Sultan's envoys to France had arrived at Busra. Even a year later Sir Sidney Smith considered it to be important that news from Egypt should reach England and India at the same time, and arranged that it should do so. In November 1799, when the French had evacuated Rome, and when Bonaparte had overthrown the Directory, at the date when we have grown accustomed to turn our eyes away from the East and to consider that the political interest of the moment was entirely European, the situation looked quite different from the point of view of Constantinople. Lord Elgin's deliberate opinion, as expressed to Lord Mornington, was that there was no chance of expelling the French from Egypt, "however able and brilliant our naval operations against them continue to be"; and that "no French settlement in Egypt could be otherwise than directed against us." If this was the deliberate opinion of well-informed Englishmen, there is nothing extravagant in Bonaparte's assumption that if he could only reduce Naples to vassalage he might yet force his way through to the sea, and succour the French troops left behind in Egypt.

Everything appeared to conspire in his favour. The Emperor Paul withdrew from the coalition, and with the Russian troops went the best part of the fighting force of Naples. The Hamiltons were recalled, and with them disappeared much of the moral support which had strengthened the queen in her resistance to France. Nelson was wanted elsewhere. England acted as if all danger was over at the moment when the greatest peril was impending.

Not so the Queen of Naples. While the dread campaign of Marengo was in preparation, she sailed from Palermo intent on seeking more energetic support from Vienna. In Leghorn she learnt the news of Marengo. More resolute than ever in the face of disaster, she turned back from Northern Italy, now once more in hostile hands, and made her way round the southern and eastern shores of the peninsula. She was landed at Trieste by a Russian captain, and took up her residence at Schönbrunn on August 18, 1800.

Unfortunately for her the foreign policy of the Austrian Court was undergoing a marked alteration under the influence of the Minister Thugut. In Thugut's opinion unconditional resistance to France was a policy no longer possible to be pursued. Thus in every direction the ground crumbled away from her feet. Only on the sea, where the armies of France could not operate, was there a gleam of hope. The long blockade of Malta ended on September 9, 1800, and the island became English soil. But small comfort could the sovereign of Naples draw from this victory, when the French had reconquered the whole of Northern and Central Italy. The advantage of possessing Malta—such as it was—accrued to England; to the Austrians Hohenlinden struck a deep and terrible blow. This decisive battle, fought just three months after the fall of Malta (December 3, 1800), pointed the arguments of Thugut unanswerably. It was idle to talk of Austria sending aid to Naples when the empire could hardly defend its own border. England could do no more than defend the coast-line of Naples, and it was clear that unless the grip of France on Northern Italy could be loosened Naples would soon be left face to face with France. The months wore anxiously and heavily on, bringing Queen Caroline nearer and nearer to the inevitable. All appeals to the Tsar were made in vain. The Emperor Paul was heart and soul on the side of the First Consul. The Peace of Lunéville, concluded on February 9, 1801, left Naples at the mercy of France; and by the Peace of Florence, concluded on March 18, 1801, Queen Caroline was to learn by bitter experience how far the last five years had removed her kingdom from the fortunate circumstances of 1796. The Peace of Florence was dictated rather than concluded. The published articles included the establishment of the kingdom of Etruria (formed partly from Neapolitan territory), the surrender of ships of war, the abandonment of Elba to France, and the closing of the ports of Naples to British and Turkish ships of war. Then came the secret and most burdensome conditions of peace. It was no longer sufficient for Bonaparte that Naples was neutral or even obedient. The South of Italy must in the future be made to serve him actively. So good an occasion for strengthening his position in the Mediterranean must be improved to the utmost. Twelve thousand French soldiers, destined to reinforce the army of the East, were therefore cantoned in Brindisi, Otranto, and Taranto, and were there to be quartered, fed, clothed and paid by Naples.

There could be no doubt about the future ; whatever might be attempted or achieved at sea, the mainland of Southern Italy was now under the heel of the First Consul. For the moment, however, the ambitions of France in the East were at an end. The victories of Abercromby and Hutchinson decided the fate of the army of Egypt. Fought contemporaneously with the conclusion of the Treaty of Florence, these battles appeared to snatch away from France the advantages acquired by the oppressive terms imposed upon Naples. The principal result of the stupendous efforts of France between March 1798 and March 1800 were, as regarded the Mediterranean, on the one hand the complete domination of Southern Italy, on the other hand the loss of the Ionian Islands and Malta. What lay before France in the future was either to reconquer these places by subduing England itself, or else to neutralize their possession by capturing Sicily.

It was understood that the occupation of Brindisi, Otranto and Taranto should last only until the conclusion of peace between France and England. A whole year elapsed before the Treaty of Florence was followed by the Treaty of Amiens (March 27, 1802). Not until the latter date was it possible, apparently, for the sovereigns of Naples to re-enter their capital. The alliance with England had maintained Ferdinand undisturbed in Palermo, but had been unable to restore him to Naples. The queen had delayed her departure from Vienna for long after the time when the Peace of Lunéville showed her that she could not hope to enlist the political sympathies of Austria. She had every reason to linger in Vienna among the associations of her childhood and to dread returning to her own exhausted kingdom. For her the welfare of her realms was a great part of the duty of her life. A true daughter of Maria Theresa, she loved the details of government, the sense of power and the sense of responsibility. It was her misfortune that her life was passed in such bewildering political circumstances. In any other epoch she would

have passed for a good, and even a great queen. The misery
of Naples was, not that the queen was weak, but that
Napoleon Bonaparte was so much too strong. As for King
Ferdinand, the nominal sovereign, he was a person of inde-
scribable triviality. He was the son of Charles the Third of
Spain, one of the greatest of Spanish kings, a man who raised
his country higher than it had stood since the death of the
Emperor Charles the Fifth. But Ferdinand of Naples inherited
not a spark of his father's capacity for governing. In the
affairs of Naples and Sicily he counted for very little. For
France, Naples and Sicily (especially Sicily) were important
posts on the road eastward, to be desired both for the immediate command of the Mediterranean conferred by their
possession, and for the indirect help they gave in adventures
further east. For England, the kingdom of Naples and Sicily
was a highly respectable part of the European system; with
its affairs the English had no concern except that, if it could
not stand alone, they must do their utmost to prevent it
falling under the control of France. Such were the views of
the principal actors in the drama, the second act of which,
that following on the truce—(rather than the peace)—of
Amiens, was now about to open.

The British Embassy in Naples is inseparably associated
with the name of Hamilton. But by the year 1803 Sir
William Hamilton had resigned, and his place had been taken
first by Drummond, and next by the First Secretary of the
Embassy—William A'Court, who afterwards became Minister
at the same Court. After a short period A'Court gave way to
the new Ambassador, Sir Gilbert Elliot's brother. This was
Sir Hugh Elliot, whose mediocre abilities were rewarded with
a grave in the north aisle of Westminster Abbey. It almost
seems to have been the policy of France to make the choice
of French Ambassadors at Naples speak out the contempt in
which the kingdom—now the vassal kingdom—was held in
Paris. First Garat; then La Combe St. Michel, a regicide;
now in 1803 the French Ambassador in Naples was another
regicide—Alquier. If the object of these appointments was

to exasperate the queen into some indiscretion, they failed.
To the Queen of Naples Bonaparte was a highly interesting
and intelligent upstart, as she not infrequently intimated to
Alquier, with all the exquisite and condescending politeness
of a great lady brought, by some odd chance, into close relations
with a personage not of her own class. Great as were the
afflictions which the queen had endured in finding herself at
the head of the buffer-state between France and England in
the Mediterranean, they were as nothing to what were to
follow in the years succeeding 1802. At first there came a
gleam of hope. The Emperor Paul had died (March 24, 1801),
and his successor was not definitely committed to the cause
of France, and soon became a bitter foe of the emperor.
From the Neapolitan point of view all the misfortunes of
Naples and Sicily came from the obstinacy with which England clung to Malta. For Neapolitans Malta was only a rock,
like any other rock. It was inconceivable to them that
England should run the risk of renewed hostilities with
France for so trifling a matter. And when it was considered
that Naples (for whose welfare the English professed so much
concern) would have to pay for English obstinacy, it is not to
be wondered at that they regarded English behaviour as not
only irregular (which from some points of view it might easily
be made out to be) but wanton and heartless.

Malta and Egypt had been shown by the action of France
to be two stepping-stones eastward in the route to India :
just as, on the alternate route, the Cape of Good Hope and
Ceylon had proved to be. Malta, to England, was therefore,
much more than a mere rock, like any other rock. It was far
too dangerous a policy for England to pursue to leave the
possession of Malta to be settled by the designedly cumbrous
provisions of the Treaty of Amiens. In England itself a
public opinion had grown up on the subject exactly resembling that which had so constantly hindered the cession of
Gibraltar. There was much expert opinion in favour of the
evacuation of Malta ; but expert opinion was powerless in
face of the expressed feeling of the country. Over the question

of Malta, then, France and England were clearly coming to blows ; equally clearly Naples would have to pay the expenses of the conflict. The first warning came in Alquier's monstrous demand that Naples should close all her ports and harbours to the ships of England—not only to ships of war, but also to merchant vessels. This took place in March 1803, under instructions from Paris.

The wretched situation of Naples was made still clearer, when Talleyrand, following the instructions of the First Consul, suggested, as a way out of the difficulty, that the occupation of Taranto by the French might run concurrently with the British occupation of Malta, and that a term of ten years should be set to both occupations by mutual agreement. In this proposal, rejected by the British Government, the rights of the sovereign of Naples were completely ignored. Still more flagrant was the treatment of the unhappy buffer-state when Gouvion St. Cyr was ordered to march 13,000 men through Neapolitan territory, and occupy all ports from Pescara to Brindisi as a material guarantee for the neutrality of Naples.

The particulars of the numerous changes of frontier and jurisdiction effected by the operations of French armies in Italy during these years have little direct historic interest. The phantom kingdom of Etruria, for example, which may one day serve as the ground of a historical romance, was called into existence, not for the purpose of obliging Spain, but for the purpose of disobliging England. The price asked from Spain in exchange for the crown of Etruria was the colony of Louisiana. The object of re-acquiring Louisiana for France was the sale of the colony to the United States of America. The sale was effected for the considerable price of three millions and a quarter sterling, a sum which was no doubt very welcome to France. But a still more desirable feature of the transaction was the obvious annoyance that it caused to England.

The ruin of Naples was determined on, not because Bonaparte cherished any particular grudge against the sister of

Marie Antoinette ; although he affected to do so in order to conceal his real intentions, and to stimulate public opinion in France in his favour. It was determined on merely as an incident in his march eastwards ; and if the renewed attempt to force his way to Egypt should be destined to the same failure as his first attempt, there would still be the consolation that he had embarrassed and perhaps bewildered England, that he had damaged her trade, and that he had proved her to be incapable of according any serious support to her allies —all of which worked to the discredit of England, and so to the advantage of France—as Bonaparte understood the interests of France. The borders and revenues of Clarke's duchy of Feltro, or of Marmont's duchy of Ragusa, or of any of the countless new principalities carved out of Italian territory for the benefit of Frenchmen, would form an interesting foot-note to the history of the period. But the bearing of their transient existence on high policy is not obvious, unless we remember Napoleon's never-relinquished plan of Eastern empire. Even supposing that the command of the sea-coast did not immediately confer the command of the sea, better times might come for France, or worse times for England. In either case it would be easier to start the armada of the East from Ragusa and Brindisi and Corfu, rather than from Toulon or Genoa. With these objects then, the western shores of the Adriatic were occupied by French troops early in 1803.

There was also the possibility that in very shame at seeing an ally of ten years' standing beggared, the English might be brought, however reluctantly, to evacuate Malta. Rage and despair filled the heart of Queen Caroline, as she saw her beloved province brought to ruin in order to gratify the senseless pride of her so-called ally. Better an open enemy than such an ally, she not unnaturally concluded. Thus the least advantage accruing from the action of the First Consul was the sowing of discord between England and Naples. For there is no doubt that England could have secured the withdrawal of French troops from Neapolitan territory at the price of resigning Malta to Bonaparte. In years gone by he had bought Corfu at the exhorbitant price of Venetia; in years to come he was to offer England far greater bribes for the possession of Sicily. Insensate and heartless as British policy must have appeared to the Queen of Naples, it was nevertheless the only true policy to pursue, the only means by which France could be held in check.

In the meantime, Naples itself and the Adriatic coast-line were occupied by the French. Acton's threat of a levée *en masse* was an idle menace, to which Alquier hardly troubled to reply; and for the rest St. Cyr was a man of admirable temper for the execution of the atrocious mission with which he was charged. He even became rather popular than otherwise at the Neapolitan Court. Unfortunately the Court and the people were not at one. Of the Royalists many would have preferred a less yielding demeanour in the face of French exactions. Of the populace, some were frankly in the French interest, some in favour of a republic, proclaiming war to the knife to the French. Every party was in strength except the English; they alone had no man's good word.

The king, excellent sportsman as he was, although a mere mockery as a sovereign, sighed for the sleepy, beautiful paganism of Sicily, where he could still hunt and fish at his ease, where nobody either talked politics or understood them, and where under the shelter of the English fleet his preserves remained silent and well-stocked, undisturbed by French armies of occupation. The fiery industry of his wife had at first diverted, then bored him; it now appeared to him futile as well as embarrassing. Acton was paid to do the work; why not leave the disagreeable interviews to him? But the queen maintained that the sovereign's presence in Naples alone prevented the Revolution from breaking out afresh, and persuaded him to stay by her side in Naples. So, in extreme discomfort and anxiety, the summer of 1803 wore on. Summer passed into autumn and winter, leaving Naples poorer every week, and more and more besieged with complaints, mostly groundless, from the French Ministry. All this time the plans for the descent on England were maturing, and in return England was busy in attempts to form a new coalition against her enemy. Russia was the most promising ally. Prussia refused to stir; Austria was still too near to Hohenlinden to think of a new enterprise against France. Naples, of course, was powerless—nay, was paying for her former friendship by enduring the most cruel exactions. The more Bonaparte saw himself threatened with an alliance between Russia and England, the more tenaciously he clung to the coasts of Naples, from which he could hold in check the Russians in Corfu and the English in Malta, the more definitely did he, according to his wont, throw all the blame of the impending outbreak upon the Queen of Naples. Assuredly if there was at this epoch in all Europe a ruler guiltless of intrigue it was Queen Maria Caroline. Nothing was more certain than that the outbreak of a new Continental war would be her ruin. The one course that could have embarrassed the French, and perhaps saved her country, was to retire to Palermo. This was the king's advice, although it was from no motive of high policy that he suggested the emigration. The withdrawal of the Court would have been the signal for popular uprisings in every direction. The scene of 1799 would have been repeated; and if the queen ceased to be Queen of Naples, she would at least have ceased to be Bonaparte's rent-collector and commissary-general, which (disguise it as she might) was all the part now left to her to play in Naples.

The exceptionally close connection of England and Naples at this epoch was the result of the geographical conditions in which Bonaparte was operating. But it was symbolized and accounted for publicly by the fact that an Englishman was Prime Minister of Naples during these eventful years. Whenever the public showed symptoms of failing interest in the wickedness of the queen, they were promptly advertised of the wickedness of her Prime Minister. Abuse of Pitt was alternated with abuse of Acton. Not unreasonably: for Acton was a remarkable man. He was destined therefore to the same fate which afterwards befell Stein, and which Bonaparte reserved for all men of originality or force of character, not excepting his own brother Louis. Mediocrity humbly waiting for the word to act—such was Bonaparte's ideal Minister. Acton did not fulfil either of these conditions, and his removal from office was accordingly decided on. The details of the quarrel which was forced upon Acton are no more interesting than the details of any similar sordid proceeding. Decorated, advanced to the rank of prince, and awarded a fine estate, Acton, overwhelmed with favours, and in enjoyment of the openly-expressed confidence of his sovereign, retired from office in May 1804, amid tributes of military and popular respect which amounted to an ovation. Mediocrities replaced him—or attempted to do so; but it was rumoured that foreign affairs were still directed from Palermo, where Acton had retired, and Alquier was proportionately anxious to see the fallen Minister settled further away from Naples. His residence at Palermo appeared to hinder the closer union with France, which could alone protect Naples from the wicked English : so ran the political cant of the last years of the Consulate. But the crushing exactions of France had already caused the pendulum to swing once more towards England. England, it was felt, had been blundering and selfish, but she had never shown any marked ill-will towards Naples. With a combination of England and Russia, it might still be possible for Naples to pay back some of the indignities which she had suffered at the hands of France.

These were dreams : 1799 had been bad ; 1803 was worse ; far worse things awaited the unhappy queen from the date when Bonaparte was transformed into the Emperor Napoleon. The exactions which the sovereigns of Naples had been compelled to endure from the hands of the First Consul left them in no mood to compliment the new emperor. The queen had rightly gauged the upstart temper of the new Charlemagne, but her wounded pride would not allow her to propitiate him ; and in addition to the political reasons for oppressing Naples, Napoleon had now an affront to avenge. Naples had neglected to pay him homage. The neutrality which the queen now alone desired for her unhappy country would assuredly never be granted. The slightest movement of Russian troops on the Cattaro, or of English troops in Malta, was urged as a sign of a secret conspiracy against which France must be protected. On December 2, 1804, Napoleon the First was crowned in Paris.

Six weeks later the queen had the pain of perusing what was, surely, the most insulting letter ever addressed to a sovereign. She was called upon to decide whether she would set Europe in flames for the sake of England, and whether she was not aware that she would be the first victim of such a conflict. If war broke out, Napoleon proceeded to assure her, her dynasty would fall, and she and hers would have to beg their bread through the streets of Europe. The rest of the letter was in the same style.

This atrocious production was unfortunately no vulgar piece of bluster. If the queen had been less sorely tried by fortune, her natural acuteness would have enabled her to see the situation as it was. In brief, whatever she did or endured Napoleon meant to have Naples. Fortunately the English alliance might be securely counted on to preserve Sicily for her dynasty. But in her natural indignation at the emperor's language she lost sight of the just outlines of the problem, and could only grasp the half-truth which was urged in so masterly a style by Napoleon. It was true, indeed, that owing to England she was about to lose Naples ; that much was obvious, and was urged with all the brutal elocution of which Napoleon was master. What should have been her consolation, and was only a source of mortification, was the reflection that if England had not existed, she would have lost not only Naples but Sicily as well. In the vast stakes of empire for which England and France were now struggling, the dual kingdom itself was but a trifle. If England won the queen would be restored, and in the meantime she would enjoy Sicily under the protection of the British fleet. If France won, Naples and Sicily became for ever fiefs of France. England was decidedly of opinion that the Queen of Naples ought to have been grateful to her. But all this was hidden from her eyes. All that she could perceive was that she was being insulted and threatened because England would not evacuate Malta—an insignificant rock with which England had no imaginable concern—and she was naturally resentful. It is not reasonable to blame her, for British ideas were far from clear. She followed her instincts of government; England followed its impulse of self-defence. But through all the years succeeding 1804 we must never lose sight of this difference of opinion: that whereas England looked on herself as the saviour of Sicily, the queen looked upon England as the ruin of Naples.

The emperor's violence was translated into diplomatic terms by Alquier. Damas (a capable soldier) must be dismissed the service; the English Ambassador must be handed his passports, and the army reduced. Naples gave way. Immediately afterwards Napoleon proposed an alliance between Eugène Beauharnais and Princess Amelia of Naples and Sicily, afterwards queen of the French. But here Queen Caroline was firm; nothing would induce her to consider such a proposal; so another grievance was added to the long list of grievances which Napoleon considered he was entitled to record against the daughter of Maria Theresa. Elliot's dismissal was the one point upon which Napoleon did not show himself inflexible. It probably did not suit his plans to commit himself unnecessarily to a diplomatic outrage.

In the meantime the third coalition against France was taking shape. Russia and Austria were its chief members; England, of course, was prominent, but inasmuch as the shock of battle would clearly be felt on land, her accession did not lessen the anxiety with which well-informed military men contemplated another conflict with France. If Napoleon could only have contrived to cross the Channel in the preceding eighteen months, many of his embarrassments would have been removed. As it was, he heard of the gathering forces with irritation, and expressed himself furiously in denunciation of the plot; the more furiously, perhaps, from the sense that the

R

new coalition would be much more easily dealt with if the fleet of England were not behind it.

As yet England alone was at war with France. The immediate object before the emperor was to strengthen his grip on Italy. Naples, it was clear, could not be further harried without precipitating matters; but the vassalage of the sole remaining state of Italy not already in recognized bondage to France was aimed at by the assumption of the Iron Crown. Even as the emperor was on his way south on this errand, Russia and England entered into alliance against him. Villeneuve, united with Gravina, had slipped through the Straits of Gibraltar, and Nelson was far away from Naples scouring the Atlantic in pursuit. Thus the impending coalition, the absence of the British fleet, and the increased authority in Italy which Napoleon would derive from his coronation in Milan, all pointed in the same direction. A ruthless and irresistible enemy, already in partial armed occupation of the country, was at the gates of Naples. He was about to assert, by the mere act of his coronation, a suzerainty over the country. It was impossible to exceed his expressed ill-will against the reigning dynasty; equally impossible, in the absence of the British fleet, to organize anything approaching resistance to his arms. At a crisis of this gravity Sir Hugh Elliot tendered to the queen the advice to despatch to Corfu what remained of the Neapolitan fleet, there to amalgamate with the sea-forces of Russia. Sicily, at all costs, must be preserved from the French; and to that end he was instructed to offer a strong British reinforcement from Malta. It is unnecessary to comment on either the wisdom or the timeliness of this counsel; for it was not a moment where any choice was possible. But the queen, as we have seen, had made up her mind. Naples was as the apple of her eye. If she was disturbed there it was because of the English; advice from Elliot, therefore, was as an enemy's counsel to her. She answered him evasively. No doubt some such attitude had been anticipated; for Elliot no longer troubled to conceal the direct and unequivocal nature of his instructions ; in the event of any further hostile move in the Mediterranean being made by France, England was to occupy Sicily.

It is impossible to combat the wisdom of these instructions. More, supposing that it was desirable in the interests of Naples to resist France to the death, it is impossible to deny their beneficence. The queen, however, cold to the English, reserved towards the French, despatched Prince Cardito to represent her at the coronation of Napoleon at Milan, and instructed him to report to her on the situation. To every other observer the assumption of the Iron Crown implied the absorption of the Italian peninsula into the French Empire, to the Queen of Naples it was a diplomatic incident to be reported on, and duly considered. Even as usefully might a bird report upon the mental attitude of the serpent to whose prey he is destined.

One must perforce pass some moments in considering the behaviour of one person, in contemplating the mighty and tangled web of interests which were involved in the conflict of France and England in the Mediterranean. It is unnecessary to enlarge on the attributes which have made the Queen of Naples the central figure of this epoch—her touching mis-fortunes, her great natural ability, her beautiful, if saddened home-life, her romantic friendship. What we must not fail to observe is that, as the sum-total of all that can be said for and against her, she entirely misapprehended the issues that were at stake, and the policy of England. All that she could perceive was that both France and England coveted her territory : France coveted Naples, England coveted Sicily. England, indeed, coveted Sicily, so that the island might be preserved for the queen, and perhaps used as the means for re-conquering Naples for her. France, on the contrary, coveted Naples in order that the queen might be despoiled of that kingdom, and of Sicily also, if that might be. But to the Queen of Naples both Powers were pirates : the one strong, the other feeble. As well as making this fundamental mistake the queen acted as if she were still living in the days of her mighty mother. She treated her despoiler with reserve and even hauteur. She omitted the common civility of sending him a decoration on the occasion of his coronation. The emperor was in that stage of a parvenu's career when petty slights rankle. Prince Cardito was treated to a scene, a public scene, such as made every other outburst of Napoleon's anger seem moderate and delicate by comparison. " Jezebel " and " Athaliah " were the least violent of his expressions, and he concluded an outburst of gutter abuse by informing the Ambassador that he would not leave the Queen of Naples ground enough to bury herself in. As usual the extravagant declamation of the emperor distilled into a formal, if exacting requisition through diplomatic channels—on this occasion into the demand for immediate recognition of himself as King of Italy. There was no time to " report " : the recognition was effected.

Surely if the queen's mind were not unhinged by misfortune this incident should have opened her eyes. Nor was this all. On July 8, 1805, Alquier informed her, without circumlocution, that the emperor designed to depose her, and was merely waiting until he had selected her successor. The queen burst into tears ; but not even the memory of this painful scene sufficed to clear her mind. From henceforth we must consider the queen as one whose ideas had no relation to the facts around her. The Russian Minister and his secret agents continued, with the best possible intentions, to arrange in conjunction with Sir Hugh Elliot and the Austrian Ambassador, masterly plans of campaign. These were based upon a joint landing from Malta and Corfu, and proceeded on the assumption that Naples could contribute 40,000 soldiers to the total force of 60,000 which the military experts deemed sufficient for a campaign against St. Cyr. It was precisely with some such plan as this in view that Napoleon had so long made a point of bleeding to death the kingdom of Naples through the agency of St. Cyr's army of occupation. The vast outlay incurred in the up-keep of this hostile force had so reduced the resources of Naples that the 40,000 armed

men existed nowhere except on the muster rolls. Armed
action by Naples against France was now an impossibility.
Queen Maria Caroline had watched the reduction of her
beautiful and prosperous kingdom to the position of a desolate
and helpless province of the French Empire. Every step in
the downward progress had been marked for her by personal
insults which she had been compelled to endure at
Napoleon's hands ; and yet she still thought it worth while to
appeal to him, as she did at this juncture, to spare her
kingdom further exactions and to allow her to resume her
attitude of neutrality. Napoleon's only reply was to order
Eugene to arrest every Neapolitan courier bound for Vienna
or St. Petersburg and to forward the despatches to Boulogne.
St. Cyr was to be reinforced with 20,000 additional men, and
to turn his army directly to the object of expelling the reigning dynasty. "The sovereigns of Naples," it was naively remarked, "will never accept our system." This was not unnatural
reluctance, seeing that the French "system" implied their
vassalage. The negotiations carried on between St. Cyr and
Alquier and Prince Cardito throughout the summer and
autumn were transparent and futile. Alquier affected indignation at the warlike attitude of Naples, when nobody knew
better than Alquier that Naples was incapable of putting an
army into the field. The Court affected ignorance of the
projected Russian and English expedition for its relief, when
that expedition had been designed under its own superintendence. St. Cyr affected alarm at the approaching invasion and
mobilized his forces, nominally to resist the Russians and
English, while he held in his hands the emperor's orders to
expel the Bourbon dynasty.

At this time England was in a thoroughly false position.
It was the fleet of England alone which rendered possible the
projected landing, for it was childish to suppose the Russians
could have faced Villeneuve alone. If the expedition failed it
was upon England alone that the Court of Naples could rely
for a refuge in Sicily, or even for the possibility of escaping
capture by the French. The contribution of soldiers promised

by England was substantial—8000 men—although smaller than the Russian force. Nevertheless, the queen had so imbued her surroundings with her insane mistrust of our policy, that Sir Hugh Elliot found himself a secondary agent in an undertaking of great magnitude, which was only feasible if he supported it. Nor could England afford to withdraw. Sicily at all costs, and Naples if possible, must be upheld. The grasp of the situation shown by Englishmen was as conspicuous as the feebleness of the policy of their allies. Sir Sidney Smith wrote, five years earlier, that the French hold on Mecca would never be shaken until their influence in the Mediterranean was destroyed. In 1805, Neapolitan statesmen were busy considering in all gravity the trifling concessions offered by Talleyrand—concessions offered simply as material to occupy their time until the emperor was ready to strike. Collingwood, on the other hand, brushed all such petty considerations aside. For him there was no doubt possible; the French obviously intended to absorb Italy, and he himself was there to save as much from them as he could. The objective of France was India, and the Mediterranean was a means to that end. All this was by now clear ; it was an elementary condition of the problem. But for England's allies, there was nothing more serious at stake than the map of Europe as it existed at the date of the Seven Years' War. On October 22, 1808, the Russians sailed from Corfu, and cast anchor in Syracuse on the 31st. The English joined them on November 7, and the allies sailed for Naples.

From the English point of view the expedition might be expensive, but could do little harm otherwise. The operations of war on land would not be decided in Italy, as England was well aware. If Napoleon was defeated in Germany all would be well, with or without English help. If he was victorious, withdrawal from Italy would become merely a question of time. But so little grasp of the situation was there at Naples, that even when the squadron was on the way, the Cabinet was divided in opinion, firstly, as to whether it should be allowed to land or not, and secondly, if it landed, whether it ought to be trusted in occupation of the strong places of the kingdom.

If anything could add to the absurdity of such a situation, it is to be remembered that Naples had not formally joined the coalition, and that Alquier was still representing the emperor at the Court of Naples; that he was required to meet as friends the Russians who were hastening to fight his master in Germany; and that Damas, formally expelled from the kingdom at Alquier's request, was summoned from Sicily to take command of the Neapolitan troops. The English were placed under the orders of Lacy, the Russian general.

At this moment came the news of Trafalgar, which momentarily decided the situation, but which was received at Naples as an incident, favourable indeed, but not of the first importance. Nor was it of immediate help to the queen, for Naples was none the less exposed to the wrath of Napoleon. More important appeared, at the time, the disembarkation of the English and Russians in the Bay of Naples on November 20. True to the last to its uncertain policy, the Government announced in the *Gazette* the "unexpected" arrival of an Anglo-Russian fleet. Alquier immediately withdrew from his Embassy and left the kingdom.

There was no plan of campaign ready. No preparations had been made to receive the troops, or to horse the batteries and the cavalry. Cantoning even so large a body of troops was no difficult matter in so great a city as Naples, and the English soon found themselves in comfortable quarters, but they showed no inclination to move. They were greeted with the usual reproaches—"haughty islanders," and so forth; but really it is hard to say why they should have shown any remarkable energy in such lukewarm company and on so foolish an errand. The key to the situation, so far as England could be of any use, was of course the hold on Sicily; a few battalions of foot and unmounted cavalry could make no great difference one way or the other when it was a question of meeting Napoleon.

Elliot, who was instructed to this effect, and who had made no secret of his instructions, now made no secret of his disapproval. He had throughout been slighted and suspected, and he instructed General Craig, commanding the forces in chief, not to allow his troops to be wasted. Naples contributed only 3000 men to the defence of its own borders, and it was at last resolved, so that the army should not remain idle, to blockade Ancona—at that time occupied by the French.

With December 1805, we reach a date when English relations to the kingdom of Naples and Sicily were to become more definite, and at the same time less agreeable, than they had been in earlier years. Up to this moment England had been the more or less trusted ally of Naples, but it had not been necessary to interfere very actively in the internal affairs of the kingdom.

When the perpetual exactions and enforced capitulations of the unhappy sovereigns had reduced their kingdom to a point where they were within measurable distance of sinking into helpless vassals of France, more definite action was imposed on England. Clearly if Naples became French, Sicily would follow the fate of Naples, failing external support. To interfere was imperative unless England was prepared to repeat the blunders of the years preceding 1797, to evacuate the Mediterranean, and leave the way open to the East. This England was resolved not to do a second time. " The Government of France never ceases to regard India as the most vulnerable part of our empire," wrote Sir Sidney Smith to Lord Keith, " and Sicily clearly must not fall into the hands of a Power with designs on India." This resolve, first arrived at by sailors, had permeated whatever there was of public opinion in existence, had overborne the earlier ideas of public men on the Mediterranean Question, and had turned insensibly into the conscious and expressed policy of England. With the consent of the Court of Naples, without it if that consent were withheld, Sicily must be occupied by England. With the rejection of this—the only policy which could be relied upon to preserve a respectable part of her authority—the ideas of the Queen of Naples ceased to have any relation to the facts of life. She was not grateful to England for saving Sicily—in her opinion England ought to have sent 100,000 men to defend Naples, or else have left her to make her peace with the emperor, oblivious of the fact that the emperor had no intention of allowing her to make her peace with him.

The Government of France at this date was conducted by Joseph Bonaparte, his brother Louis and Cambacérès. To them Gallo was still accredited from the Court of Naples. Gallo had never been under any illusions as to the helplessness of Naples in the presence of French exactions. His one object had been to make the vassalage of his native country (since vassalage was its obvious destiny) as little burdensome as possible. The landing of Russians and English he looked on as a suicidal act for Naples. Whatever course might have been profitable, to welcome an armed force was sheer lunacy. It could effect nothing, could only exasperate the emperor, and as regarded Gallo himself, it placed him in the worst possible position, unless he frankly threw over the sovereigns who had been guilty of such imprudence. As he was fond of public life, and conscious of sufficient abilities, he was by no means prepared to be ruined by his attachment to an impossible cause. His adherence to the new dynasty has been so often blamed that it seems only fair to point out the egregious folly which forced him away from his ancient allegiance. The King of Naples was not a person for whom it is possible to cherish any high feelings of respect or regard. But he possessed the simple wisdom of the entirely selfish person—washed his hands of all connection with the proposed aggressive action against France, and pursued his pleasures with an undisturbed mind. Whatever might be right the queen's action was clearly wrong, and he would have nothing to do with it. The Crown Prince with a heavy heart allowed himself to be dragged, an unwilling agent, in the course of his mother's headstrong policy. But after Austerlitz, Elliot and Craig made no pretence of being interested in the campaign, and busied themselves with securing the transport necessary for the impending retreat to Sicily.

That which Collingwood had openly spoken out, that which all observers of Mediterranean affairs had long recognized as axiomatic where France was concerned, that which was hidden from the eyes of Queen Caroline alone, now befell Naples. The Bourbon dynasty was declared to be deposed, and the throne was conferred upon Joseph Bonaparte. Joseph was despatched to Naples as the emperor's representative, but without the royal title for the moment. The English general declined to sacrifice brave lives in a hopeless resistance to the Imperial armies under Masséna and St. Cyr. But he was alone in his opinion, the Russian generals advocating resistance, while the queen ordered the guns of Sicily to fire on any English ships approaching the coast with the view of effecting a landing. The senseless and wicked abuse with which the queen has been loaded has long been recognized as baseless, but in view of the insane order to fire on the English, an order which cut off her last chance of escaping capture by the emperor, it is difficult to maintain that her reason was unclouded by misfortune.

The Emperor Alexander settled the matter for the moment by ordering the immediate withdrawal of the Russian forces. Naples, left face to face with France, now offered unconditional surrender, and the queen herself wrote to the Emperor Napoleon, offering to expel the English, to place the French in possession of the strong places of Naples, to surrender her fleet, and finally to abdicate, to procure her husband's abdication, and to give over what was left of the governing power into the hands of her son. The queen preferred to write this to Napoleon rather than to accept the secure shelter of the English fleet.

We have travelled far from the days of Nelson. What had caused this astonishing change in the queen's mind? In the absence of any evidence of particular affronts or disagreeables endured by her at the hands of the English, we can only conclude that the large policy forced on England in defence of the empire was beyond her comprehension, and that the only feature of that policy which touched her nearly—the occupation of Sicily—was unintelligible to her except as an act of piracy. This much we must conclude, unless we decide that her reason was clouded by misfortune.

Her abject appeal received no reply. Ambassador after ambassador went in vain from Naples to Rome, where Cardinal Fesch represented the Imperial authority. The Russians and English proceeded steadily with the work of embarkation.

Idle schemes for the defence of Naples were put forward in the Council of State; but in the private apartments of the Royal palace the servants were busy packing. By mid-January, 1806, the French were once more in Rome; Joseph Bonaparte arrived a few days after Masséna, bearing the strictest injunctions to hold no communications with the Court of Naples, and to leave the queen's letters unanswered. Not even this intolerable insolence could turn the heart of the queen from her enemy.

Naples, then, was to be French; but in Napoleon's eyes Sicily was the better half of Naples. On January 31 he was already impatient. " Lose not a moment in seizing Sicily," he wrote to his brother, adding that even the naval difficulties might be overcome if advantage were immediately taken of the enemy's bewilderment. Donzelot, afterwards Governor-General of the Ionian Islands, was selected for the Sicilian command, and Lamarque was joined with him.

At this juncture we must commend the simple wisdom of the king. He had no intention of enjoying the hospitality of the Emperor Napoleon. All that he could hope for from France was a pension and a moderate establishment on more or less humiliating terms. He could get as much as that, or more, from the English. He took no exaggerated view of the English policy or promises. He did not suppose that they were putting themselves to the vast inconvenience of defending Sicily out of a Quixotic attachment to his person. But he did believe that in their own interests they would protect the island from the French. Even supposing the worst said of them to be true—and in his opinion it was nowhere near the truth—he would still find in Sicily all that he wanted. A river or two, some good sea-fishing, the Royal preserves, a comfortable hunting-box, and Ferdinand of Naples asked nothing more. It was useless to argue any longer with the queen ; so the day after Joseph entered Rome Ferdinand sailed for Palermo. The queen delayed more than a fortnight, conducting what she called "negotiations" with the lord of forty thousand veterans, against whom she could still muster the Royal Guard, six hundred strong. On February 11, she sailed from the Neapolitan capital, and two days later accepted the hospitality of the detested English. The next day the Regency received the French troops in state on their entry into the capital. Gaëta alone held out under the Prince of Hesse-Phillipstal. Exactly a fortnight had passed since the emperor's public announcement that the dynasty had ceased to reign.

The Royal family had escaped, penniless. Obviously some one must pay their expenses. The resistance to France was still maintained in the southern provinces of the mainland ; obviously it must be supported. Naturally England was applied to. As regarded resistance, England had had enough of armed expeditions for the moment ; but the provision made by England for the maintenance of the Royal family in exile was £400,000 a year. The mainland appeared to be entirely subjugated with the exception of Gaëta. The kingdom of the Two Sicilies was declared to be an integral part of the French Empire ; but the new sovereign was not yet formally announced. Gallo enthusiastically supported the new state of things ; not unreasonably was his adhesion taken to imply that French dominion was sincerely welcomed by the former officers of state. Somewhat ominous was the lightning rapidity with which England had contrived to close the ports of Naples to all possibility of trade. Not Napoleon himself was more rapid in his movements than that wonderful fleet. Before even the new king was crowned, it became clear that it would be no easy matter to find him a revenue.

In studying these events from a point of time separated from their occurrence by nearly a century, it would be natural to expect that the fleet which could do such great things was conscious of its power. It would not be surprising to find that whatever else might be obscure, the policy of extending the coast-line of the French Empire so as to include the Tyrrhenian and the Adriatic, left no doubt in the minds of contemporaries as to the course which was imposed upon England. Obviously the fleet must close on the (now) hostile shores of Italy. That was done; but the effect of so doing—the concentration of all marine activity on the Italian coasts—was not immediately apparent. On the contrary, in accordance with the purblind traditions of the Continental war as understood by England, there was immediately fitted out a series of those unfortunate expeditions which were intended to menace the power of the French Empire, and only succeeded in bringing the fighting power of the British Empire into complete contempt. This ought not to be forgotten when contemporary opinion on the Peninsular campaign is under consideration. The Peninsular War was a campaign which, at the outset, was not in appearance more likely to be productive of good results than the Buenos Ayres expedition of General Whitelocke, or the Egyptian expedition of General Fraser. The Buenos Ayres expedition was planned with the view of preventing France obtaining a hold on the colonies of Spain, and the Egyptian expedition was planned, apparently in oblivion of the change in Mediterranean affairs which had been wrought by the battle of Trafalgar, and the extension of the coast-line of the French Empire. Even in 1807, it was still thought desirable to forestall Napoleon in Egypt, although it is difficult to understand how he was to repeat the manœuvres of 1798 without a fleet.

In the meantime, Joseph's task was one which might have daunted even a great military organizer; and Joseph was a man who—whether as prince, as general, or as a private individual—was always found wanting. The policy of France for the past ten years had been to weaken Naples by unceasing requisitions, and by the quartering of a large army on the resources of the country. That army was now more than doubled, while the resources of the country had been systematically ruined—with a definite object. Naples—naturally a fertile land, and formerly not too much oppressed—was now a wilderness. , It was not, perhaps, in more evil case than Prussia when Frederick upheld his throne against Russia and Austria; but Joseph possessed no fraction of Frederick's vigour and administrative capacity. He did his best, however; appointed Gallo to be his Minister for Foreign Affairs, and other Neapolitan nobles to his council. The only other Neapolitan whose name we need remember is Campochiaro, who was appointed Minister of what we should call Crown Lands and Household. A sinister choice was made when Salicetti was forced on Joseph as Neapolitan Minister of Police. Salicetti was a veritable Scarpia. Formerly a member of the Convention, a friend of Paoli from the moment when Paoli turned upon his benefactors, possessed of Corsican unscrupulousness and Corsican guile, but destitute of Corsican fire, Salicetti made himself and his master detested. One other Frenchman—if a Corsican is a Frenchman—was included in Joseph's Cabinet—Miot, the Minister of War: a good man. So, with all the blame thrown upon his shoulders if anything went wrong, and with all the glory of the kingdom of Naples illuminating his brother's crown, the new King of Naples entered upon his sorry reign. Miot was capable; Joseph himself was conscientious and kindly; it was the activity of Salicetti, and the baleful interference of Napoleon, which made the Bonaparte dynasty detested. It is remarkable to find Napoleon incessantly urging his brother to hang, shoot, ravage and oppress. These are the resources of a sovereign who is weak, and must needs make the terror of his name do the work of many regiments. Such measures have nothing to do with the work of a dynasty which is intended to be permanent, and which is supported from the outset by an overwhelming military force.

"At bottom," it was said of Napoleon the Third, "his

Italian blood despises the Gauls." Whether or not this was true of Napoleon the Third, it was certainly not true of his uncle. On the contrary, he viewed the Italian populations with contempt. The Neapolitans were, for him, "canaille" —and worthless. Every man found with a poniard in his possession ought to be shot. All the lazzaroni should be hunted into the hills to perish.

The French occupation of Naples was at best little more than a dangerous experiment. But the spirit which informed these savage comments, and which found vent in the activity of Salicetti, made the experiment hopeless from the outset. Joseph's reign, if conducted on Joseph's principles, need not have been irksome. He had even some measure of that attractive manner that counts so much for men in high place. "You call me a stranger," he said to the notables whom he met on tour, "and in part I am ; that is, I am only half Italian. But what was your late ruling family? Not even that; for in so far as it was not French it was Spanish; where-as I am at least half an Italian." This was modest and persuasive. The fatal weakness of Joseph's position was that it rested on the support of the French army of occupation, and the emperor was not the man to lend forty thousand men without exacting the most punctilious deference to his commands in return. Moreover, wherever King Joseph's line of march brought him near the sea, he was made to feel another limitation of his authority; for if he did not actually sight British vessels he heard news of them. His very coronation, which took place in May 1806, was marred by the sudden appearance of the Mediterranean Squadron in the Bay of Naples. No shots were fired ; but the silent menace was characteristic of the grim humour of Sir Sidney Smith.

The news that Naples had found a new king reached Queen Caroline in Palermo, together with the news that Gallo's example had been widely followed, and that in the new Court of Naples few faces were changed except the sovereigns'. The Neapolitan nobility thronged round King Joseph, and added lustre to his brilliant Court. Caroline (at least, at first) held no Court in Palermo. Seven years had passed since she had taken refuge there under circumstances which at the time had appeared hard; but in the light of the events of 1806 the flight with the Hamiltons in Nelson's ship seemed happy by comparison. Indeed, for a reigning sovereign, her plight was terrible. Not only were lord chamberlains and masters of the household wanting, but even nurses and housemaids. Even life itself, and such peace of mind as may remain to an exiled queen, she owed to the detested English, so that life and repose were themselves poisoned in the miserable mind of Caroline of Naples.

Sir Hugh Elliot was still accredited to her Court; the command of the Mediterranean Squadron was held by Collingwood. Both officers held the same simple instruction; viz. to uphold the authority of the Royal family, and to keep the French out of Sicily at any cost. The king was heart and soul on the side of England. His absorbing selfishness might possibly have blinded him; but Acton's opinion is conclusive, if any doubts could possibly exist on the subject, of the imperative need of relying wholly upon England at this juncture. For Acton was entirely devoted to the interests of the Royal house; and in so far as those interests could be said to be divided, he had hitherto been rather in the queen's interests than in the king's.

Powerless at sea, Napoleon was still in the plenitude of his authority over the mainland of Europe. The dukedom of Dalmatia marked the extension of the French Empire down the eastern shore of the Adriatic. To this Russia and England replied by blockading the ports of what had been hitherto Austrian territory; so that the situation in the Tyrrhenian Sea was repeated in the Adriatic; the French all-powerful on shore, their enemies easily dominating the waterway.

Corfu, soon to pass from Russian to French hands, commanded the Adriatic, and threatened Soult's duchy. The queen welcomed with open arms the suggestion that Russia should occupy Sicily as well as Corfu; but the king and

Acton, with greater wisdom as we should hold, refused their consent, and preferred to trust exclusively to England. Whether the wisdom was Acton's or the king's, it implied a confiding temper; for the British army of occupation was in trifling force—all told, it numbered under 8000 men—and the nightmare of an invasion by the French in rowing-boats haunted the queen and her advisers. To all their other crimes the English had now added a new offence—they were wantonly risking the safety of Sicily by refusing the help of Russia, and then playing into the hands of the French. But England had recently passed through the crisis of the camp at Boulogne; and was well aware that if Napoleon Bonaparte could not effect the passage of the Channel, it was in the last degree unlikely that Joseph Bonaparte would effect the passage of the Straits of Messina. Many months must pass before even the fleet of rowing-boats could be got ready; for the nucleus of the fleet of invasion was nothing more than one frigate, one corvette, and some pleasure-boats. The English took full advantage of the situation. It was a waste of time and resources to pour troops into Sicily when England could carry the war into the enemy's country. Disregarding the queen's protests and lamentations, the English proceeded to further her interests in their own way. Gaëta still held out. On May 12, the day following King Joseph's coronation, the British occupied Capri. The French garrison marched out with the honours of war, and was succeeded by an English garrison. The British occupation of Capri inaugurated a well-reasoned and successful policy. It was that of so distracting and harassing the French army of occupation, that it should not only be prevented from attacking Sicily, but should be materially hampered in its operations on the mainland. By way of strengthening Joseph's hold upon his new kingdom, Napoleon could devise no more useful measure than punitive expeditions. Punitive expeditions may be valuable expedients when we are dealing with hill-robbers over whom we have no permanent authority; but to conduct punitive expeditions within one's own borders is surely not a very statesmanlike

s

measure. Desaix had burnt villages on the Nile as a punitive measure; and this at a time when Egypt was nominally a French colony. The same measure was followed by Pelissier in Algiers. But it is not possible to imagine the English burning Belgaon or Satara, because a seditious plot had been hatched, or a British official murdered there.

Joseph, then, under orders from Paris, was conciliating the affections of his subjects by burning their villages at the time when his obvious incapacity to protect his own coast-line was made clear by the British occupation of Capri. The incidents are typical.

In the meantime the French continued their progress on the mainland. Ragusa, afterwards erected into a duchy for Marmont, was occupied by the end of May, and immediately harried by land and sea by the Russians. Every fresh extension of the coast-line gave the enemies of France a new point of attack. England threatened Castellamare and Torre dell'Annunziata, and seized Ischia and Procida. It was already clear that there would be no peace for Naples whilst England held Sicily. As no means existed, for the moment, for expelling them by force, the emperor opened negotiations on the subject, and made his first offer. Ferdinand was to abandon Sicily, and receive in compensation the Hanseatic Towns in sovereignty, guaranteed by England. Neither Ferdinand nor Napoleon were under any illusion. By both it was clearly understood that Ferdinand, as a sovereign, existed merely on sufferance. Ferdinand bore us no grudges for a situation which we had not created; and, as a matter of history, was perfectly prepared to retire to England if the protection of Sicily should prove to be beyond our power. The offer of the Hanseatic Towns was therefore made to England, and by England rejected. It was immediately followed by the offer of Albania, together with the two duchies of Ragusa and Dalmatia. This offer was rejected also; and the rejection was accompanied by the intimation that the evacuation of Sicily was a point on which England could not consent to negotiate. A very natural attempt to

confuse the situation, by making Russia a third party to the
negotiation, failed; and amid gathering wrath at Paris, and
increasing anxiety at Naples, the year 1807 wore on. For
the emperor, Sicily had become the one important spot in
Europe, the position on the possession of which hinged the
peace of the world. Masséna was chosen for the Sicilian
expedition; but the Straits of Messina remained to be
crossed. No doubt the emperor's estimate of 15,000 men as
a force sufficient for the subjugation of the island was correct;
but the problem was to ferry 15,000 men across the straits in
the face of a hostile fleet; and although Napoleon never
wearied of urging his brother to the attack of Sicily, he made
no attempt to solve the main problem, beyond saying that it
presented no difficulties. Yet he must surely have remembered
the camp at Boulogne.

Joseph was no soldier. His brother told him so, with
Napoleonic frankness. But the frankness inflicted no wound;
for Joseph knew his own military helplessness better even than
his brother. The bent of his mind was solely towards kind
deeds and simple enjoyments; so that to the inherent difficulties
of the situation there was added the embarrassment that the
sovereign had no mind to face them. The internal troubles
of the kingdom, concealed by the strong rule of Queen Maria
Caroline, were more conspicuous than ever. They were
unimaginable to a reader unacquainted with the social conditions of pre-Revolutionary Europe; but they do not affect
the course of this narrative, except in so far as they weakened
the offensive force of the French kingdom of Naples. Indirectly, and with the lapse of time, the administrative im-
provements introduced by Joseph would have strengthened
the central Government considerably. But the emperor was
not thinking of the state of Naples as it might be in 1830;
but of what it was in 1806, and how far the resources of the
kingdom might be made to subserve his own ambitions; and
in Napoleon's opinion, Joseph was wasting his time.

In the meantime, Prince Louis of Hesse-Phillipstal held the
fortress of Gaëta for King Ferdinand, Sir Sidney Smith

supporting him at sea. At first, in the turmoil of the overthrow of the Bourbons, Gaëta was hardly remarked. After
the establishment of King Joseph, it was observed that one
spot in Naples was as yet unconquered. By degrees the
fortress of Gaëta became the point on which the eyes of
Italians were fixed. The continual repulses of the French,
the irrepressible vigour of the prince, and the news of the
supplies poured in by Sir Sidney Smith, aroused the
enthusiasm of every man who cared for Italy. Nearly 10,000
of the best troops of France were engaged in the attack, which
was directed by Masséna himself, with no less capable subordinates than Lamarque, Donzelot, and Gardanne. It
seemed at first as if the task was beyond the French.
Amantea was captured with a garrison of 400 men in the
beginning of July, while at the same time Sir John Stuart,
commanding in Sicily, landed nearly 6000 men in Calabria for
his famous campaign of Maida. On July 4, 1806, Reynier,
who commanded in the south, attacked the British lines, with
too great confidence in the terror of the French arms and the
worthlessness of the Neapolitans. Out of an attacking force
of nearly 5000 men, he lost nearly 1000 in killed, wounded
and prisoners, and was compelled to fall back upon Catanzaro.
The not very severe loss of between 300 and 400 on the
British side indicated a short but keen struggle, and did not
materially mar an important little victory.

Joseph appealed to his brother for reinforcements. With
40,000 men under his command, he had been unable, in six
months, either to keep the peace in his kingdom, or to capture
Gaëta, or to avert a notable defeat of one of his principal
generals. It must be admitted that the English were perform-
ing all, and more than all, that they had promised. Verdier and
Reynier were ordered to evacuate Catanzaro and concentrate
on Cassano. Prince Eugene was directed to march from the
North of Italy to protect the Adriatic coast, and relieve the
hard-pressed armies of the South and Centre. Clearly, if
Gaëta were not captured, the cause of the French kingdom of
Naples would be seriously imperilled. The tide set steadily

against France. Ponza and Ventotiene fell; and a marked coolness towards the French on the part of Neapolitans was the consequence. Queen Caroline burst out into bitter reproaches against the English for not pursuing the advantage, and replacing her upon the throne of Naples. The English, the king and Acton, all saw clearly enough that the advantages gained were precisely of the kind that were not to be pursued. There was never the smallest intention on the part of the British to expel king Joseph : not even when, after Stuart's victory, Reynier was compelled to evacuate a long and important piece of Neapolitan territory. So that the English were doing their utmost, and really achieving marked success, in the cause of a queen who never ceased raving at them for not attempting the impossible.

If raving had been all the queen could achieve, her lamentations would have excited pity, but could not have altered the policy of England. But Queen Caroline was still powerful. There existed a Sicilian Parliament at this epoch. It was not a vigorous body ; but it met occasionally, voted and tendered advice to the sovereign. As the occupation of Sicily promised to be of indefinite duration, it was, from the outset, an object of that occupation to make the island prosperous, to open it up to trade, and to raise perhaps some permanent revenue, or at least to develop some permanent sources of supply. The Parliament was the obvious instrument to be used for this desirable end. Instantly the queen ranged herself against Acton, summoned around her the nobility—whose privileges, she asserted, were threatened by the projected measures of internal reform—and had organized a violent opposition before even Acton had formulated a governmental policy. To assert, as was earlier ventured, that the queen's ideas had ceased to have any relation to the facts of life, seems to understate the exaltation of her mental attitude on this and numerous similar occasions.

A pledge of good faith to the Royal family was now given in the landing of 5000 British troops to strengthen the garrison of Sicily. It would be supposed that the queen, who had been reproaching the English for leaving the island exposed to an invasion in rowing-boats, would have breathed a sigh of relief at the arrival of the tardy reinforcements. On the contrary, forgetful of her earlier anxieties, she ordered the troops to the mainland forthwith. But the troops were not placed under her orders; and although she could and did show her animus against England by forbidding the men to be quartered in Palermo, that was the extent of her power of interference. The troops were cantoned at different spots on the northern and eastern coasts, and directed to watch movements on the mainland. Her orders to them to march through Calabria were ignored; and the relief of Gaëta was attempted instead, though it was attempted too late. On July 18, 1806, the famous fortress avoided the assault of Masséna by surrendering. The garrison was handsomely treated, and allowed to re-embark for Sicily.

By August 1806 the state of things on the mainland was clearly defined. Within the lines of King Joseph's army the authority of King Joseph was recognized. Within the lines of the British army, which had not yet retreated to Sicily, the authority of Queen Caroline was recognized. Over the rest of the country there was no authority existing: every man defended himself as best he might against his neighbour. Stuart in vain sought to bring order into the country-side. He put a price upon the head of a notorious brigand who responded by putting the same price on Stuart's. The one reward had as good a chance of being earned as the other. There is nothing in such a wild state of things that justifies the parallel afterwards drawn between Italian disorder, and the popular uprising in Spain against the French. To say this is not to say that the Italians were unpatriotic. It is only to say that a very large proportion of Italian restlessness was, at this epoch, mere impatience of any government at all. It could not be relied upon to rally round any cause because its impulse was anarchic. The liberators of the next generation sprang from a very different class from the mountaineers of Calabria. Quite rightly did England refuse to risk any considerable enterprise on the mainland where the only " popular " support it was likely to receive was of so fickle and feverish a nature.

The fall of Gaëta set free an army for what was practically the re-conquest of Calabria. The defence had done its work. None of these temporary occupations of Italian ports and islands was intended to be permament. Their object was delay. The result of the defence of Gaëta was that ten thousand men had been occupied for four months, and that opportunity was thereby afforded for the campaign of Maida and the rising in Calabria. These were adequate results.

The capture of Reggio and Scilla by the British, involving a further loss to King Joseph of 1000 men—the strength of the captured garrisons—marked the high-tide of the insurgents' success. The Sicilian Court, ever too eager, had selected new governors for the conquered districts. But although the flag of Ferdinand was actually hoisted as far north as Catanzaro, it was obvious that the movement must collapse if Masséna took its suppression seriously in hand. For the moment Reynier was fighting for his life. His retreat, bravely conducted, involved heavy losses amounting to hundreds of prisoners. His pathway was marked by burning villages and ruined homesteads. Wherever he touched the seas the watchful English cruisers cannonaded his flank. Wherever he touched the hills sharp-shooters crouched behind every rock. In the plains the people would barricade their houses, to which Reynier responded by sacking them of all useful provisions and burning the fabrics of the buildings to the ground.

August 1806 saw Masséna join hands with Reynier and commence his avenging march southwards. It was a march of plunder and fire and slaughter. The troops, to whom full licence was given, did not need the savage invective of the emperor to spur them on to the welcome task. The commonplace asperities of Reynier paled before the barbaric rush of Masséna. Outlawry and confiscation was the portion of all who had goods to be seized and who were still at large themselves.

To this fearful onslaught the English could offer little opposition. Continuing the irritating plan of campaign on which they had hitherto relied, they occupied Procida and Ischia. But amid the horrors of the campaign on the mainland these little successes passed unnoticed. The French army pushed on, relentless, irresistible, and by the end of September the rebellion was crushed. The net results of the six months' campaign were an impartial loathing of both French and English, implanted in the hearts of the Southern Italians, the loss of thousands in killed, wounded and prisoners, on the part of the French, the retention by the English of numerous posts on the coast-line, and the extension of French authority to the Straits of Messina—where they were at liberty to contemplate anew the island of Sicily, which had been the object of the whole campaign, and from the conquest of which, apparently, they were as far removed as ever. Incidentally the whole of Calabria was ruined and the population infuriated against their conquerors. Obviously supplies could neither be drawn from the country nor forwarded through the country : an important consideration. The cause of Queen Caroline of Naples was strengthened by the difficulty of keeping an army in the field against her.

Russia, reluctant, but with experimental views, had at last been drawn into negotiation on "the Sicilian Question." A new solution was offered by the emperor. The Hanseatic Towns and Albanian coast having been rejected by the Sicilian Royal family, as compensation for the surrender of the island, the offer of the Balearic Islands (Spanish ground) was now made to the Crown Prince. With Russia some progress could be made on these lines : with England none whatever. It was possible to negotiate over the Ionian Islands, whose somewhat farcical " independence " was recognized on both sides—an " independence " qualified by the retention of a Russian garrison of 4000 men in Corfu. There were also Franco-Russian transactions in respect of Ragusa, and the mouth of the Cattaro. The cause of Ferdinand was vaguely described as " abandoned." But when the turn of

England came, all these mutual concessions were found to weigh but little in the face of the Sicilian Question. Napoleon was resolved on the acquisition of Sicily for his brother. England was equally resolute that under no conditions whatever should Sicily pass into the hands of the French. By strenuous efforts in St. Petersburg the Russian concessions were made to depend on the accession of England to the Tripartite Treaty : whereupon all fell through ; and the great war of 1806–1807 broke out. " I will never lay down arms," wrote the emperor to his brother, " till Sicily is yours."

The dramatic collapse of the Prussian kingdom in October 1806 strengthened Joseph's throne for the moment. Without seeing very clearly into the question, the waverers in Naples could appreciate the importance of keeping on good terms with the victor of Jena and Auerstadt ; although those crushing blows did not obviously make it easier to cross the Straits of Messina without boats.

To the difficulty of expelling the English and suppressing the revolt, there succeeded the difficulty of dealing with typhus and malaria—doughtier foes than poniards and bayonets. It became more and more difficult to provision Ragusa. Turkey, however, came to the assistance of France, and closed the Dardanelles to the Russian fleet advancing from the north to reinforce the Adriatic Squadron.

We now come to two incidents of Mediterranean history, both of which are intimately bound up with the story of the British occupation of Sicily. These are the forcing of the Dardanelles by Admiral Duckworth, and the invasion of Egypt by General Fraser. Of these the second is entirely forgotten, and the first is remembered in history as an episode, eccentric, inexplicable ; and memorable, chiefly, if not entirely, as a kind of moral lesson enforcing humility.

In point of fact, they formed part of the threefold scheme of England for heading Napoleon off from the East. The first measure of precaution was to hold Sicily : the second to seize the Turkish fleet : the third to garrison Egypt.

A writer of genius has, in our own days, demonstrated in so

masterly a manner the influence of sea-power upon history that we have accustomed ourselves to a view of military matters hardly more sober than our earlier opinion on naval matters. It is scarcely too much to say that we have reduced Mediterranean, if not European, history to one word—Nelson. Nelson himself with all his vanity—and his vanity was gigantic, like his genius—would have hesitated to make such a claim. It may, indeed, be sound ; but history gives us some reasons for pausing before we accept it altogether. The battle of the Nile was said to have decided the course of events. Yet five years after the fleet of France was annihilated, Nelson found himself face to face with a naval Power superior to his own. The people of this country were in the utmost alarm at the impending invasion from France ; and England was, in appearance, in an even worse position than if the battle of the Nile had never been fought.

Trafalgar followed ; and even more decisive language has been used about Trafalgar than about the battle of the Nile. Within a year of Trafalgar, England was in even greater alarm for the Indian Empire than in 1797. Fox held Sicily in considerable force ; a large army was despatched for the occupation of Egypt ; a strong naval force was employed to force the Dardanelles and seize the Turkish fleet. It is strange that such efforts should be called for if Trafalgar really settled the situation. Clearly there is something in warfare to be considered as well as the command of the sea. The recovery of France after 1798 was achieved by the simple process of rebuilding the French fleet and adding the Spanish fleet to the French fleet. The menacing situation of the year succeeding Trafalgar was brought about by the achievements of Napoleon on the mainland : achievements which had practically brought the land frontier of France within striking distance of Constantinople. The place of the Spanish fleet was to be taken by the Turkish fleet, and the descent on Egypt to be attempted from Constantinople instead of from Toulon, as in 1798.

Contemporary observers did not rest in the conclusion that

the command of the sea secured England from all risks. On the contrary they held that it was counterbalanced, if not out-weighed, by the fact that all the ports of both of the shores of Italy were in the hands of the French. Contemporary observers did not look on the possession of Egypt as a question of small interest to England. On the contrary, they were of opinion that Egypt was the key to India, and they employed that expression as governing all views of Mediterranean politics which could make any claim to sanity. They may well have been wrong ; but such were their views.

Sir Hugh Elliot's despatch to Lord Collingwood dated June 1, 1806, set forth in grave and restrained language what were the elements of the situation with which he found himself face to face. The great successes of the French in Germany had enabled the emperor to put any pressure he pleased upon the Austrian monarchy. That commanding position was being used in order to secure the passage of French troops through whatever Austrian territory intervened between Northern Italy and the Balkan Peninsula. The subjugation of both coasts of Italy and the domination of the western coast-line of the Balkan Peninsula gave him innumer-able advantages for the furthering of the expedition. His object was clear : to expel the Russians from Dalmatia, and to seize the European possessions of the Sultan of Turkey. Constantinople, which Napoleon had approached from Acre in 1800, was now to be approached from the Cattaro, and when seized was to be used as the point of departure for Alex-andria. Sicily, if possible, was to be seized in the meanwhile.

On March 20, 1807, Alexandria was occupied by the troops detached from Sicily. For the better appreciation of the campaign which followed, it may be useful to review shortly the internal condition of Egypt at the moment of the invasion. The Mameluke Beys continued to regard the country as their own ; but they were weakened by the assault of Napoleon and by their own internal dissensions. The authority of the Sultan, although less phantasmal than in the year 1798, was not exercised with sufficient vigour to justify the name of a government. This was fortunate, perhaps ; for the solid interests of the country, the merchants, great and small, the agricultural interest—everybody, in fact, who had anything to lose—were opposed to government by the Turks. The desert tribes were closing in upon Lower Egypt, and according to the information furnished to us by Major Missett the country might fairly be described as derelict. Major Missett resided in Egypt after the withdrawal of the British troops in 1803, and supplied very clear and sound information to his chiefs. But even Major Missett was completely deceived as to the future of Egypt. According to him, the solid interests of the country would be in favour, firstly, of an English occupation, secondly (if England could not interfere), of a French occupation ; lastly, and most reluctantly, they would resign themselves to government from Constantinople, but only if no other form of government could possibly be devised.

At this crisis nobody appears to have thought of Mehemet Ali, the nominal representative of the Sultan, who became Viceroy of Egypt, after many vicissitudes, in the summer of 1805. From the English point of view Mehemet Ali was merely the successor of the many powerless functionaries who had so long enjoyed a dignified position at the tolerance of the Beys. In so far as anything else was known about Mehemet Ali, England was aware that he was heart and soul in the interest of the French, through whom he had obtained his appointment to Egypt. No longer after Trafalgar than June 1806, Egypt was considered as "constantly exposed to the danger of invasion" by France. With Mehemet Ali as viceroy, the "invasion" might well, for all that could be foreseen, be made an eagerly accepted invitation. From the point of view of British interests, from the point of view of the interests of the people and country of Egypt, or as a move in the large enterprise of baffling Napoleon, the occupation of Egypt in 1807 must therefore be considered—like so many other incidents in the Mediterranean—as a measure designed rather to preventing France securing an advantage than to securing one for England.

The expedition promised well. It was, in fact, more or less called for. The "second and long meditated invasion of Egypt" by the French had been expected for the last two years, and was believed to be imminent when Mehemet Ali was appointed viceroy in succession to Khurshid Pacha. Mehemet had, been in a somewhat inferior command in Candia, and his re-appearance in a great post was supposed to be the keystone of the fabric of French authority in the East. The real aims of Mehemet Ali were not suspected ; still less was his capacity realized. During the previous five years, while speculations of all kinds had been rife, the viceroy was the last person to attract attention. Not unnaturally—in the absence of a commanding figure—Egypt was looked on as the most alarming centre of intrigue and agitation in the whole of the Mediterranean. Lord Hobart, writing in May 1801, held that nothing could prevent a renewal of the French invasion, except England holding the country in force. Lord Hutchinson, avoiding speculation, merely pointed out that Egypt must inevitably fall a prey to some European Power before long. " The chief use of Malta " (wrote Major Missett) "is to guarantee us Egypt." " If Turkey regains control," said the same authority, " Egypt remains ' an easy prey to the first invading Power.' " No doubt all this was as clearly realized by M. Lesseps, the French Consul, as by Major Missett ; and General Stuart—Lord Hutchinson's successor—recorded that the French had never for a moment relaxed their intention of re-conquering the country. Lord Elgin, our Ambassador at Constantinople, writing from Buyukdere on December 21, 1802, when Sebastiani had already been two months in the country, contributed an interesting essay on the situation ; but his advice hardly amounted to more than an opinion that something might be done with the Beys. The final evacuation of Egypt by England had taken place on March 11, 1803, and the troops arrived at Malta on the 27th and 28th of the same month. Sebastiani, bearing the title, at once ambiguous and menacing, of the First Consul's plenipotentiary to the Levant, had already published

(January 30) the outrageous report on the French re-conquest of Egypt, in which country he had landed in the month of October 1802.

It is hardly matter for surprise that England kept a firm hold upon Malta and Sicily. As regards Egypt, the opinion of every officer who was qualified to have an opinion was that if England were not in Egypt, France would be there. Nevertheless, the country had been evacuated in 1803. Even four years later, when the expedition of General Fraser was equipped, it operated without confidence, on the defensive, and with no clear ideas on the work before it. Whether this confusion and hesitation arose from jarring counsels, or from fear of responsibility, or from the fatality which brought England into ridicule at Belleisle in 1800, and Buenos Ayres in 1806, confusion and hesitation did undoubtedly preside over the operations of 1807 in the Levant. Small wonder if, on safer ground, England persisted in the policy adopted since the commencement of the Revolutionary wars. Whatever might happen in Egypt, England could not be other than stronger for her hold on Sicily and Malta. But the very magnitude of the reasons which appeared so cogent to English eyes, obscured them to the view of Queen Caroline of Naples. To her, Europe was still the Europe of Maria Theresa. A policy which embraced the fate of two, and perhaps three continents, and in which her historic kingdom was of but fractional importance, was unintelligible to her. So far as she was capable of judging, England had sacrificed Naples by clinging to Malta, and was now about to use Sicily as a stepping-stone to the conquest of Egypt.

Regarding the Beys as the important factors of Egyptian politics, Major Missett informed them that England was invading Egypt for the purpose of baffling the French, not in order to hold the country for England. The language was somewhat inflated in view of the course of the campaign. Colonel Wauchope, detached for the occupation of Rosetta, was surprised and slain, together with four hundred men under his command. A second attempt to secure this outpost resulted in the capture of Colonel Macleod and seven hundred more prisoners. The martial Beys were little likely to respect an invader who displayed so small a knowledge of the elements of warfare in a strange country. In point of fact, Fraser's instructions laid it down that the possession of Alexandria was "the exclusive and only object" of the expedition. But the imperious necessity of gathering in food and forage for the invading force called for these unfortunate enterprises.

Mehemet Ali, watching passing events with the eye of a governing genius, used these disasters to enlist on his side the sympathies of the Arabs. He had early in his Egyptian career decided that the Beys were impracticable people, and he made in consequence no serious bid for their support. As regards the English, he observed their invasion without animosity. Their obvious incapacity was agreeable to him ; and he rather welcomed them than otherwise. Although a nominee of the French, he had no intention of becoming their tool, and the best way of making use of the French, without allowing them to make use of him, was clearly to avoid quarrelling with England. "We shall not long retain this key to our East Indian possessions," wrote Missett to Sir Alexander Ball on May 5, 1807.

Mehemet Ali, having watched the demeanour of the English, and having found that it corresponded with the instructions reported to have been furnished to General Fraser, and actually announced by that officer as governing his movements, proceeded to make an offer. Instead of seeking to restore the broken authority of the Mamelukes, why should not the English rather ally themselves with the Pacha ? Fraser's instructions would not allow him to enter into negotiations on this basis ; and the first hint of his inevitable withdrawal (if he did not make peace with Mehemet Ali) was given in his despatch to Windham on May 19, 1807. The question was (so he reported) whether it would be possible to fortify Alexandria. The commanding engineer, Captain (afterwards F.M. Sir John) Burgoyne, decided in the negative. The army was now reduced to four thousand effectives. Of these he might count on three thousand for the defence of Alexandria ; and considering the uncertain temper of the Arabs, and the great extent and bad condition of the fortifications, it was out of the question to attempt the defence. After two months, Major Missett reported to Lord Castlereagh (July 23, 1807) that the British must either be reinforced or withdraw at once. Mehemet Ali, with consummate prudence, had let the invaders severely alone. He was determined to avoid coming to blows if possible ; and he relied upon ophthalmia and fever, lack of supplies, and the reluctance to engage in heavy and useless expense to do the work for him. On August 30, Fraser received his orders to retire. The orders came from Sir John Moore, who had replaced Fox in the Sicilian command, and on September 5, 1807, the terms of the evacuation were signed at Dámanhúr. Mehemet Ali behaved with the large wisdom that might have been expected from him. The Egyptians who had sided with the English were amnestied ; his treatment of them was described as " most fair and liberal," his treatment of the English who had fallen into his hands as " truly noble." On September 19, the inglorious campaign of six months' duration was concluded by withdrawal to Messina. The campaign of 1807 cannot be said to have been quite barren of results, if it revealed the extent of Mehemet's capacity, and brought home the con- viction that in his hands Egypt could take care of itself. A repetition of the invasion of 1798 would clearly be a task of extreme difficulty. " The Mamelukes as a political body may now be considered extinct," wrote Missett to Lord Castlereagh on October 21, 1807.

We have now to consider the naval operations which were carried on contemporaneously with Fraser's expedition to Egypt.

" Some late proceedings on the part of the Turkish Govern- ment, indicating the increasing influence of the French in their councils, and a disposition in the Porte to abandon the alliance which has happily subsisted between that Government and His

Majesty"—so run the opening lines of the "most secret" instructions to Admiral Duckworth, dated January 13, 1807—" the
getting possession, and next to that, the destruction of the
Turkish fleet, is the object of the first consideration." It is to
be remembered that these proceedings, as well as the Egyptian
expedition, had been taken in the face of a state of public
affairs from which the direct influence of Napoleon had long
been withdrawn. The emperor was far away in the east and
north of Europe, and the disturbed state of the Mediterranean
was owing to the impulse of the emperor's will acting through
many and sometimes reluctant agents. What might happen
if Napoleon once more turned his face southwards, and began
to operate in the favourable conditions of 1807, and how he
could best be faced, was a problem which engaged the anxious
attention of the Cabinet of 1807.

Duckworth's instructions are significant. He was ordered
to provision and water for four months at Gibraltar, pick up his
squadron on the way, and make his way to Constantinople,
where he was to take the offensive on the following lines : he
was to demand the immediate surrender of the Turkish fleet,
together with a sufficient equipment of military stores and
supplies for the public arsenals ; failing instant compliance
with these terms, he was to lay Constantinople in ashes. No
negotiation was to last more than half-an-hour.

Rarely have such peremptory orders issued from Whitehall;
they measure, perhaps, the intensity of the feeling in England
on the question of the Mediterranean and the East.

The fiasco was dramatic in its completeness. The fleet
failed to approach within several miles of the Turkish forts ;
and was compelled to retire to avoid a severe handling, which,
after all, it did not altogether escape.

The battle of Jena had strengthened French prestige in
Egypt ; the battle of Eylau (fought while Duckworth was on
his way from Gibraltar to Constantinople) was a stepping-stone
to the battle of Friedland, fought on June 14. The Peace of
Tilsit followed on July 9, and material modification of the state
of affairs in the Mediterranean ensued.

T

In the meantime, however, Queen Caroline had determined to reconquer Naples with her own troops, since the English would not help her. Remembering the campaign of 1799, she entrusted the command of her forces to a man whose name might be expected to work on the imagination of the populace. The hero of Gaëta was to do the work of Cardinal Ruffo. So far she calculated well; but to make no allowance for the presence of seventy-five thousand seasoned regulars of the French army in the kingdom of Naples was a serious oversight. Naturally the complete rout and partial destruction of the invading force took place as soon as Reynier came face to face with it. So ill-judged an enterprise made no difference in the situation either way.

But very different was the effect of the Treaty of Tilsit on the politics of the Mediterranean. Firstly, King Joseph was recognized as King of Naples by Russia ; secondly, Russia evacuated the Cattaro ; and thirdly, Russia withdrew the garrison from the Ionian Islands, and resigned them in full sovereignty to France. It is unnecessary to point out how greatly these concessions strengthened the hands of Joseph Bonaparte. But there were two counter-considerations. The first and most important was that Dalmatia was by now exhausted. No army could subsist there, far less manœuvre in the country, without regular supplies from Italy. Obviously to move on Constantinople overland was not so easy as it had appeared to be a year before. The second counter-consideration was, that the heavy military expenses of the emperor's campaigns did not allow of his increasing his navy, so as to keep pace with the growth of the British navy. Napoleon himself was perfectly aware of the importance of Corfu. He even went so far as to say that it was, for the moment, the most important place in Europe.

But by abandoning the plan of moving through Dalmatia in favour of the plan of operating through Corfu, he in reality played directly into the hands of England. The English could not interfere with him on land ; the time was rapidly drawing near when he could make no pretence of interfering with the English at sea. Consequently, although the provisions of the Treaty of Tilsit appeared so menacing to England at the time, it is rather from 1807 that must be dated the set of the tide against France, until it left England in 1815 masters of the Mediterranean. No such ambition filled the minds of the Cabinet at the time. Briefly this was concluded : that if, on the one hand, Napoleon could never hold the Ionian Islands, and would certainly not conquer Sicily, on the other hand, it was in the last degree unlikely that England would ever be able to expel the Bonapartes from Naples. Nothing better could be expected than to accept this conclusion, and to make the best of Sicily for the benefit of the House of Bourbon. This was a sensible resolve. It brought steadiness to the councils of England—councils which were somewhat discredited by the wild adventures and painful failures of Fraser and Duckworth.

But what satisfied England was only the beginning of Queen Caroline's discontent. August 1806 had seen the final removal of Acton from the councils of his sovereign, and the recall of Sir Hugh Elliot, the Minister who had so often incurred her displeasure by his lack of sympathy with her views. Elliot was succeeded by General Fox, who was no more in sympathy with her than Elliot had been ; Acton was succeeded by Circello, a creature of her own. The internal government of the island was thus set in sharp opposition to the ideas of the real managers of the affairs of the island—the directors of the English occupation. Not all the misfortunes of Caroline had been able to teach her that the English were her only friends—the only friends, at least, that could be of any service to her.

The changes in the British appointments in Sicily were numerous. Fox was replaced in the civil command by Sir William Drummond in the spring of 1807 ; he resigned the military command-in-chief to General Stuart in the winter of the same year. The English policy underwent no change ; but the repeated removal of important officials gave it an appearance of vacillation which strengthened the queen's party.

The queen's lamentations—reproaches for not attempting the impossible—were silently endured. She, on her side, was by no means reconciled to the presence of the British, still less to their measures. In extreme discomfort on both sides, but with some obvious improvement in affairs when considered from the British point of view, the year 1808 opened. This was the year of the Spanish enterprise of Napoleon.

This enterprise, of which the deposition of the House of Braganza was but the opening stage, was in every way disastrous for the French Emperor. But one result, not hitherto fully recognized, of turning the arms of France in the direction of Spain, was a great and immediate relief of the tension in British affairs in the Mediterranean. What England was still fearing, when 1808 opened, was that Napoleon might himself take in hand the programme sketched by Sir Hugh Elliot in 1806, and march upon Constantinople overland. The fact that the Sultan called himself the ally of France, was a consideration. Another may have been Napoleon's belief in striking effects; the overthrow of two historic thrones offered dramatic possibilities which were not to be met on the road to Constantinople. The infraction of the Berlin Decrees had to be punished, and the enforcement of the Milan Decrees secured. So the opportunity passed away.

The increase of Joseph's authority in this year was considerable. The kingdom of Etruria and much of the states of the Church were added to his dominions. He was now the sovereign of a more considerable state than the Two Sicilies had ever been. The recapture of Reggio and Scilla (two of those numerous posts on the Italian mainland, whose occupation by England had contributed so much to Joseph's uneasiness), early in 1808, had considerable effect. These small successes proved that if Sicily was for the English, the mainland was no less definitely for Joseph. The situation became more clearly defined by the fall of Reggio and Scilla, and yet more clearly defined by the failure of Admiral Gantheaume's naval enterprise at the same time. The event made little impression in comparison with the sensation caused by the sudden elevation of King Joseph to the Spanish throne. June 1808 saw Joseph already announced as King of Spain. Jourdan was left behind in military command in Naples; but no regency of the kingdom was appointed for the short interregnum. The apparent effect of these transactions was to give to France the undisputed command of the Mediterranean by securing to her the control of the entire coast-line. The actual effect of adding the coast-line of Spain to the already too widely extended coast-line of the French Empire, was to increase the opportunities for offensive naval action by England, and to throw extra burdens upon the overtaxed energies of the army of France. The abandonment of the plan for grasping the Balkan Peninsula—a plan which could hardly have been hindered by England—was a decided set-back to the fortunes of France. From the first moment of the reign of the new King of Naples, it must be remembered that he was fighting a losing battle. If Naples itself was in a more settled condition in 1808 than it was in 1806, on the other hand, the fortunes of France, with which those of the Bonaparte dynasty of Naples were intimately bound up, were steadily on the decline. Ninety thousand Frenchmen lay in garrison within the kingdom of Naples. Before the end of the Spanish War 300,000 Frenchmen were occupied in Spain and Portugal wasting the energies and depleting the resources of the French Empire. Nothing could have been more favourable to England than the Spanish undertaking. It is obvious that as regards the politics of the Mediterranean, the force occupied in vain struggles in Spain—or even one-half of that force—thrown into the Balkan Peninsula would have created a most serious situation for England. She could not have seriously hindered the operations of such an expedition. Even if Sicily had offered a convenient base—as it would have done—the very great distance of any possible point of attack in the Balkan Peninsula would have proved a serious impediment, while Naples and its huge garrison would have formed an invaluable base and recruiting-ground for the French.

No such alarming situation lay before England. There was nothing more serious to deal with than the hostile energies of Queen Caroline, with whom, undoubtedly, it was in vain to cope, except through the medium of a will stronger by far than her own. The need of a more vigorous tone in British dealings with the Sicilian Court produced another diplomatic move. Sir William Drummond was recalled; and Lord Amherst succeeded him.

Separated, as we are, by nearly a century of time from the events of 1808, it is easy to point out how strong was the position of England when once the tide of war had receded from the North of Italy, and the armies of France had become entangled in Spain. It was not so understood at the time. It was true that Napoleon was engaged in an enterprise which left him no time to think of Italy; but it was not certain that he would be long occupied there. There was much talk of the popular uprising which would destroy his armies; but there had been fierce enough popular risings in Italy, and they had been suppressed without leaving any permanent mark upon the face of the country. It was true that England had declared herself not to be at war with the Spanish people; and that she was generally understood to be preparing to interfere in force in the affairs of the Spanish peninsula. But England had already interfered in force in the affairs of the Italian peninsula; and even success—considerable in its way—had hardly hindered the advance of the French armies. There was nothing in these reflections which justified a sanguine view of the cause of England in the Mediterranean in the year 1808. The character of the new king made the situation appear to be decidedly alarming. Joachim of Naples, formerly Grand Duke of Berg, the consort of Caroline Bonaparte, was now forty years of age. He was, perhaps, the best cavalry leader of his own or any other time. He had many of the qualities that make a man successful, and even a good sovereign. He was industrious and capable, loved publicity and public life; and in all that he did and said, in the atmosphere of daring or gallantry with which he had surrounded his name, he gave proof of that intense and joyous vitality which at once charms and bends the wills of men. That he was popular with his new subjects is not to say much; for the Neapolitans had twice welcomed Ferdinand of Bourbon with transports of delight, after expelling him with menace and execration. They had accorded to King Joseph precisely the same ovation as they now accorded to King Joachim; and they even transferred to the new queen the title, affectionately deferential, which they had in days gone by conferred on the saddened old lady who was breaking her heart in Palermo. Caroline of Habsburg had been beautiful thirty years before, and her new subjects had acclaimed her "La bella Carolina!" Caroline Murat, née Bonaparte, was radiantly beautiful in 1808, and was also acclaimed "La bella Carolina" by the Neapolitans. It did not count for much then, that King Joachim was received with shouts of enthusiasm, with joyous peals and a popular ovation. What gave cause, and just cause, for fearing the new king was his boundless energy and remarkable capacity. Not three weeks after his entry into Naples he had wrested Capri from the English. The island surrendered on October 16th, 1808, having been held by England since April 13th, 1806—just two years and a half. The siege was not a little remarkable. Under cover of night, and in deepest secrecy, the expedition was prepared: it embarked and landed. The omnipresent fleet was soon encircling the place and seeking out weak spots in the French lines. But the weak spots were few; the French had fortified and entrenched themselves with great rapidity and thoroughness. They had contrived to provision themselves for two months; so that the British garrison was compelled to surrender under the eyes of the British fleet.

A success so considerable and striking, within the first month of the new king's reign, showed how great was the difference between King Joachim and the diffident King Joseph. It had been impossible to feel any enthusiasm for

King Joseph. But it was quite possible to feel enthusiasm, and even to risk something, for the sake of King Joachim; and of this there was soon abundant evidence in the disturbed state of public opinion in Sicily, and particularly in Messina.

King Joachim might have become really dangerous to England, but for the conduct of the Emperor Napoleon. The essential baseness of the emperor's character was nowhere more abundantly displayed than in his dealings with those on whom he lavished his favours. His revolting suspiciousness, his total disbelief in disinterested or honourable conduct, his hateful mania for interference on every possible occasion, the grossness of his language, and his all-powerful position in Europe made the lot of his subordinates hardly endurable. The smallest sign of originality or initiative aroused his spleen. Murat had both originality and initiative; and he was, therefore, early marked out for a series of humiliations, all more or less public, which nullified much of his influence.

While Murat was laying the foundation of a great Italian kingdom, a kingdom which he sanguinely supposed he would be allowed to build up, Napoleon was busy with his entry into Madrid. At the same time he was urging his Ministers to equip and despatch an expedition for the conquest of Sicily: the island was hardly ever out of his mind. But the naval resources of France were inadequate for operations in the face of the British Mediterranean Squadron: if only Sicilian internal discontent and sympathy with Murat could be suppressed, there was nothing to fear for Sicily. Undoubtedly, if Napoleon could have brought himself to trust his brother-in-law, that sympathy and discontent would have set the whole island aflame within a year, and would have caused the withdrawal of England without further efforts on the part of France.

Hardly had the emperor completed, as he imagined, the subjugation of Spain, when his attention was claimed by the new breach with Austria. The effects on the politics of

Sicily was that the queen, an Austrian by birth, found herself sympathizing with the invader of her country, and the Russians, nominally allies of France, showed a marked leaning towards a British alliance. The Russians were not slow to note the humiliating position into which any Power sank that was rash enough to " ally " itself with France : they had no desire to see their Tsar in the position of King Joachim.

The Austrian campaign began ; and for a brief period succeeding the glorious battles of Aspern and Essling (May 21 and 22, 1809), it appeared as if, for the first time, Napoleon had found his match. The immediate result in the Mediterranean was the renewal of the British attack on the mainland, which resulted in the capture of Ischia and Procida. This was a return blow for the loss of Capri, eight months earlier. But now, as always in Italian affairs, the English could only play an auxiliary part : the larger operations were decided elsewhere—on this occasion on the field of Wagram. Ischia and Procida were evacuated at the end of July 1809, amid torrents of abuse and reproaches from the Court of Palermo. But England had accustomed herself to neglect the outcries of the queen, and to rest content with an effective, if somewhat undignified plan of campaign. This is to say the worst that can be said of it : to say the best, is to say that the British fleet was omnipresent ; that the operations in the case of Capri or Scilla were repeated in the case of innumerable smaller coast defences ; that no convoy of provisions was safe from their cruisers ; that forts were blown up, communications cut, and serious losses daily inflicted upon the French, without giving them the chance of striking a return blow ; that the naval campaign on the Italian coasts was one of the most useful, if not the most brilliant, of the countless services rendered by the navy to the country.

The treaty following on the campaign of Wagram cut Austria off from the Adriatic altogether, put Trieste into the hands of France, isolated Sicily, and made the land operations of French armies in the north of Italy easier than ever. So considerable an accession of strength to the cause of French authority in the Peninsula was of good omen to King Joachim. Unfortunately, the events of the last twenty years had given him a kingdom which could only be maintained by the show of overwhelming force. Such a force was undoubtedly present; no repetition of Cardinal Ruffo's exploit of 1799, was possible in 1809. But the feelings—not yet called by the name of " national "—to which the Cardinal had appealed, still prevailed. They might be stilled, but they could not be stifled ; and they found expression in the rise of a secret society, which was to give trouble to King Joachim, and which was to give more than trouble to many a government, long after King Joachim was dead—the Carbonari.

The marriage of the Emperor Napoleon with the Archduchess Maria Louisa took place in March 1810; and as regards Italy it had the double effect of connecting King Joachim with the House of Habsburg, and of connecting Queen Caroline still more closely with the House of Bonaparte. The queen had always admired Napoleon, had constantly been accused of carrying on correspondence with him, and was now connected with him by marriage. These facts were not without their weight in the troubled times that were soon to befall Sicily. When the queen's conduct is under discussion, however, it ought to be remembered that while she hated England with a deadly hatred, she had no confidence in the emperor. When she is charged with intriguing with Napoleon against England, it ought not to be forgotten that she placed it on record that she anticipated the fate of the Duc d'Enghien for herself if ever she fell into the emperor's hands. The simple, if vague old word " distraught," alone describes the state of mind of a lady who could hate as profoundly, and for so long, her friends as well as her oppressors.

The marriage of the emperor had been preceded by another marriage, which, though less celebrated at the time, came to have more important results to Europe than the startling alliance of Napoleon. The wandering and impoverished heir of Egalité Orléans had visited, among other Courts, the Court of Queen Caroline at Palermo. The remembrance of the Princess Amélie of Bourbon had drawn him to Palermo again in the autumn of the year 1809, when in extreme poverty, and with only the reluctant consent of his wife's mother, the future king of the French was married to Princess Amélie. The prince's remarkable tenacity and powers of negotiation were thus actively engaged on the side of Queen Caroline and her family. Before five years were past very important consequences were to flow from this alliance. The immediate results of the presence of Louis Philippe in Palermo was that the senseless carping and fault-finding with the British Government ceased almost entirely. Whatever else Louis Philippe may have been, he was a monument of common-sense. Unfortunately he was supposed to possess military talent as well; and he was despatched to Spain in order to take over a command of patriotic irregulars. This work, for which he was quite unfitted, withdrew him from Palermo, where he might have been of the greatest use to the English as well as to the Court.

In the meantime the English had made some way in the island. Whatever could be done with money they achieved. Road-making went on apace. Fortresses were repaired, and batteries mounted. That which Napoleon had feared in 1806 had come to pass. At that date he had urged his brother to seize Sicily by a *coup de main* in the first confusion of the flight of the Royal family from Naples. If time were granted, he had argued, the attack on Sicily would prove more and more difficult every year. Four years had passed since Sicily virtually passed into British hands. Everywhere were to be seen signs of reviving trade, renewed confidence, and a more settled state of public affairs. As regarded armaments, England was able to dispense with a large force. With no more than 7000 men all told, she was content to await the assault of the King of Naples, whose army could not have been less than ten times as numerous.

Joachim, King of Naples, and grand admiral of the French Empire, was now about to attack the English in their own element. But the days were gone when a brilliant cavalry leader could assume, without fear of discomfiture, the duties of a sea-captain. Fortunately the king committed himself to no large enterprise. He marched his armies to the coast, and assembled transports opposite Messina. But beyond manœuvring the small boats more or less unsuccessfully, he could venture nothing. Sicily was not Capri, to be taken by a night surprise. Moreover, the great distance of the king's camp from his capital tempted him to have supplies forwarded by sea, instead of trusting them to the chances of bad roads and venturesome banditti. By long practice the British fleet had attained an almost miraculous agility in detecting and following up convoys. When the ships were too numerous to be towed into a harbour covered by British cannon they were sunk. Something approaching a scarcity prevailed in the gay camp of the King of Naples: the capture of Sicily seemed further off than ever.

Nearly 30,000 men were assembled on the northern side of the Straits of Messina. Transports to the number of 500 were kept in waiting to convey the troops across. All through the summer of 1810, the French, who had contrived to bring up some heavy guns, bombarded the English lines across the Straits, and even made one attempt to cross. This was the work of General Cavaignac, who was driven back with the loss of about 1000 men.

So the whole of the year was wasted, when in October 1810 the king broke up his camp and returned to Naples. King Joachim's view of the situation had from the first been sound and statesmanlike, even crafty. He had recognized that the affairs of Naples called for all his energies and all his talents. He prepared to devote his whole time to easing the burdens of his subjects, and to making himself so popular personally, that the Sicilians should begin to regret their separation from a kingdom which had long been so nearly associated with them, and which was now so agreeable a place of residence.

He had proposed to allow the logic of facts to appeal to the Sicilians for him, until their impatience at the intrusion of a foreign garrison should lead to an insurrection. There was no imaginable reason why an army of foreigners (heretics, too) should be quartered in Sicily—no reason that could appeal to a Sicilian, the most conservative and least inquisitive of men. The Sicilian of 1810 knew no more about Eastern politics, or the designs of Napoleon, than he did about the politics of Mars. What he could appreciate, and was rapidly coming to appreciate, was the advantage of the island being governed well. In brief, Murat had determined to fight British good government in Sicily with still better government in Naples; and this the emperor was determined that he should not do. Napoleon, understanding only one medium of authority—force —admitted no principle of government except terror. He therefore commanded the enterprise of 1810, with the result that the kingdom of Naples was weakened by the failure, the English were still more firmly rooted in Sicily, and a whole valuable year was wasted. Worse than this, the kingdom of Naples was thrown back into all the old disorders—disorder of finance, consequent on the vast and fruitless expenditure of the Messina campaign; disorder of the country-side, consequent on arrested trade and the ever-growing impoverishment of the people.

To make the 'contrast between Naples and Sicily redound to the credit of the island government was a question of introducing a few of the elementary appliances of civilization. The island was fairly rich in natural products. To enable the inhabitants to exchange their wares, bridges and roads alone were needed. The authority of the central government was unshaken; probably there was less brigandage in Sicily than in Naples. But torture was still freely used in criminal investigations, and there was no certainty about judicial procedure. The historic island still possessed a constitution, which dated from the Middle Ages. It comprised three estates, of which the first was the ecclesiastical. Three archbishops, seven bishops and fifty-one mitred abbots represented the Church in a constitution where the nobles with seats only numbered 227. By the side of this imposing House of Lords, the Commons— representatives of free towns for the most part—mustered only forty-three strong. The nobles possessed all but sovereign authority on their estates, and even before the period of British domination in the island, the Crown asserted itself with difficulty against its more powerful feudatories. The tendency of British control was, as everywhere, to encourage the commercial classes. The influence of the nobles, who lived a great deal on their estates, was the daily pressure of men whose position was unchallenged, and who were the only visible sources of authority. The Commons had no influence except when the House met ; this took place every four years. The clergy were omnipresent and all-powerful. When the House was not sitting it is not to be supposed that the Crown was absolute. A standing committee of the three estates with advisory functions resided in Palermo. The authority of the House had a genuine basis in the control which it possessed over the distribution of taxation. The king's civil list, granted by the estates, amounted in 1810 to £200,000 a year ; but he was expected to maintain his land and sea forces out of this sum, and to keep up his Court as well. His private estates were very valuable : otherwise it is difficult to see how he could avoid indebtedness, even after receiving what appears at first sight to be a considerable grant.

Indebtedness might have been avoided with care ; but as neither king nor queen paid the smallest attention to money matters, and exacted no accounts from the Minister, the finances of the realm were from the outset in inextricable confusion. The sovereigns regarded their confusion—in so far as they took any notice of it—with a serenity which has been described as Royal; a less dignified adjective would be more appropriate in the circumstances.

The contribution of England to the needs of the Royal family was a subsidy of £400,000 a year, payable under treaty obligations. Although in our own days £600,000 could not be a very substantial contribution to the expenses of a campaign, warfare in Sicily in 1810 was not carried on on a very extravagant scale. The "fleet" of Naples, which the king was supposed to keep up, was a mere skeleton force : the British fleet—not the Neapolitan—kept the French at bay. At the present time it is commonly calculated that every soldier landed in India costs his Government £100 before he is landed. At this rate, £600,000 would not go far ; but the simple habits of the Sicilian made calculations of this kind altogether out of scale. Besides this income in sterling the king could count on untouched and incalculable resources in food and material from the Royal ponds, rivers, preserves, grazing-grounds and forests. It would be too much to expect from Ferdinand of Naples the smallest display of energy in the public service. But with even common honesty, or the simple resolve to apply money given him for certain purposes to these purposes and none other, Ferdinand could have kept on foot a considerable force of armed men. The queen was far above her consort in force of character; but together with many good, and some great qualities, she possessed, unfortunately, some which were neither good nor great. She insisted on keeping up a Court at Palermo which would not have disgraced Versailles in the days of her sister Marie Antoinette. She flung money right and left without any sense of responsibility. The Court was even more embarrassed when it was drawing £600,000 a year than it had been when it was drawing nothing. This may have been "Royal indifference" to many ; but the English, who had to bear crushing taxes in order to carry on the war, begrudged their £400,000 squandered in Palermo in order to give effect to Queen Caroline's ideas of her Royal dignity. When, in addition, we remember that the queen's party was strengthened by her command of much ready-money, and that the queen worked undisguisedly against England, it will be admitted that something very like an impossible situation had been created. With every desire to do justice to both sides it is difficult to assign any share of blame to England. Up to 1811 there had been no false step, unless over-reliance upon the word of the sovereign was a false step, as it certainly was one which cost dear.

Elsewhere a strange state of things arose. The English garrison was perfectly well-behaved ; pay was regularly remitted, and as regularly spent by the soldiers. The officers maintained the social traditions (never economical) of the British army. Messina prospered exceedingly in such generous hands. The Sicilians came to ask each other why the only place in the island where men laid by money and lived easily happened to be the places occupied by a foreign garrison. Undoubtedly the population was still loyal to the reigning family : it was an object to keep their loyalty unimpaired. Nevertheless the action of the king and queen, and the good behaviour of the troops, was producing precisely the opposite effect. The vast expenditure of the Court gave rise to much bitterness. " So much money could not be honestly come by," was the poor folk's comment. The nobles found that the easy selfishness which they had indulged for centuries was no longer looked on admiringly by their tenants. The natural questions were asked : why the nobles who drew so much from the country did nothing in return ; while the English who were pouring money into the country were never weary of cleaning, draining and road-building. King Joachim had feared that unless his own government was famous for its soundness, he would never be able to win the hearts of the Sicilians. The worst he feared was coming to pass : not only were his own hands tied, but the Sicilians were coming to have something like an affection for those whom it was the fashion to describe as their conquerors and oppressors. There was a public opinion—not strong yet, but steadily growing—in favour of annexation to England. The Cabinet was guiltless of attempting to foster such an idea ; it is idle abuse, worthy only of the Court of Palermo, to pretend anything of the kind. But the report reached Queen Caroline, highly coloured no doubt, and drove that unhappy lady to a frenzy of wrath and hatred.

However far the public opinion in favour of a closer

connection with England may have gone, and however it may
have originated, there could be no doubt about the changed
attitude of the Sicilian Parliament. That usually docile body
met in 1810 after the usual four years' recess, and showed
itself most disagreeably hostile to the plans for providing the
Court with more money for its extravagance. Far from
showing itself accommodating, the Parliament showed a decided
tendency to look into accounts and to require redress of
grievances, and a marked reluctance to vote anything like the
sums demanded by the Court.

With all this England was not concerned. She had,
throughout, scrupulously kept clear (not without difficulty)
of all invitations to influence local politics, or to interfere
in the internal affairs of the kingdom. But in the quarrelsome temper of the Court, these new embarrassments were
all ascribed to external influence. Nothing could be more
ridiculous. In protesting against the exactions of the king
and queen the lead was taken, not by popular representatives
with whose ignorance English agents might be supposed to
have tampered, but by the Neapolitan nobility, who made no
secret of disliking the English and all their ways.

These agitations were all thoroughly unwelcome to the
English. All that was wanted was that the island might be
kept reasonably quiet until the struggle with Napoleon was
over. The quarrels of the king with the queen, the quarrels
of both with the nobles, the angry protests of the merchants,
the sullen attitude of the peasantry were all equally alarming
and annoying to the British Ambassador and the British
general. Something more than annoyance was felt when the
queen dispensed with the consent of Parliament, and proceeded
to levy what taxes she pleased by force of Royal proclamation.
With acquiescence in this abuse of authority England must
have resigned herself to playing so sorry a part in Sicilian
affairs, that no Cabinet could have ventured to face the House
after its acquiescence had become known. The situation was
most anxious. The House of Bourbon must undoubtedly be
maintained in Sicily, and maintained by England, because

U

England alone could maintain it. Not that the English people had any longer an affection for the House of Bourbon —Queen Caroline had cured them of that—but because the island must certainly not fall to the Bonapartes, and because even if the rule of the Bourbons was unsound, it certainly was no part of the duty of England to provide the island with a better government. But it was equally no part of her duty to stand by and see tyranny and oppression exercised under the British flag, and subsidized from the resources of the United Kingdom.

To accept the situation, to allow unconstitutional government to be forced upon an unwilling people under the protection of British ships of war, would have been to violate every tradition of English national life. To allow England to be worried out of the island, would have been to open the door to France. Within a month King Joachim would have been incontestably king of the Two Sicilies. In this dilemma there was but one step to take : a representative at the Court of Palermo must be selected who would bring home to the headstrong queen some sense of her dependence upon England, and the need of deferring to English ideas of what was fitting in the first principles of government. This most ungrateful task was imposed upon Lord William Bentinck, who was appointed to succeed Lord Amherst in the spring of 1811. Lord William presented his credentials on July 25, 1811, and took over the military command of the land forces from General Stuart ten days later. In order to give additional weight to his remonstrances, the new Ambassador had been appointed commander-in-chief, as well as Ambassador, and was also granted the command of the sea forces. Malta alone was withdrawn from his authority. Here England was represented by a man as masterful as Bentinck himself, but with a far wider grasp of public affairs—Thomas Maitland. He was thus clothed with an authority which for variety and effect was probably unique ; it was, at least, highly exceptional. There is no contesting the wisdom of the measure. The stakes for which England was playing were too high for it to be possible to run any further risks. In the year 1811, if England lost Sicily, Malta would go next. With Malta would go the command of the Mediterranean. After that, Egypt must pass to France, with all the consequences to the Indian Empire which had been foretold in the gloomy year 1798. These vaticinations may have left on one side some important considerations ; they may have been fundamentally unsound. But with the experience gained in twenty years of warfare, no contemporary observer could be blamed for accepting them as only too well founded. They were supported by judgments as varied as the minds of Nelson and Hugh Elliot, Collingwood and Sidney Smith, and they were daily repeated in the observations of travellers and residents, official and unofficial, in the Mediterranean.

To avert the catastrophe, England must cling to Sicily ; and to ensure that end the choice of Lord William Bentinck may be said to have been, on the whole, a wise choice. He was in the prime of life, being thirty-six years of age, and full of vigour and determination. There never was a man upon whom responsibility rested more lightly. He had military experience, having made a campaign under Suvarov, and commanded a brigade in Spain ; and he held the rank of lieutenant-general in the British army. He was deeply imbued with the principles of the Revolution of 1688. His very name was a guarantee that Queen Caroline would not repeat, with impunity, the oppressive acts of Stuart misgovernment. There were some counter-considerations. Lord William had held high civil employment before, and he had failed conspicuously. As Governor of Madras he had involved himself and the East India Company in embarrassments of the most alarming nature. It was undeniable that he had the high sense of public duty, which was the proud birthright of the Whig aristocracy. But together with the capacity for hard work, and the determination to carry through what he believed to be sound measures of government, Lord William also possessed, in only too full measure, the arrogance of the Whig noble. A total incapacity to take any point of view except his own, and a considerable endowment of what, for want of a better word, one must rest content to call sheer stupidity, made up a character from which anything might be expected. For the moment, the situation called for vigour and determination ; and Lord William might always be counted upon for both these qualities.

The queen had taken advantage of the interregnum at the British Embassy, to arrest and banish the leading nobles who had protested against the imposition of new taxes by Royal proclamation. With inconceivable infatuation she had seized and expelled them from the island, under the very guns of the British fleet. Here was a challenge of an unmistakable nature thrown down to the new Ambassador. Queen Caroline, by open and striking acts of tyranny, laid claim to the fullest exercise of absolute power in Sicily. Among the evil consequences of this wild action was the confirmation it afforded of the secret understanding between Napoleon and Queen Caroline. The new French Empress was Queen Caroline's grand-daughter. It was firmly believed in the island that the queen had bought her peace with her grand-daughter's husband, at the price of expelling the English from Sicily. If this was untrue, nobody was more to blame for the wide-spread belief in the fiction than the queen herself. At the time, the expulsion of five nobles who took the side of the English, was regarded as proof positive of the queen's secret understanding with the emperor. Acton, whose vast experience might have been of some use to the new Ambassador, died on August 12, 1811, and Queen Caroline was left face to face with Lord William Bentinck. There was no possibility of mistaking Lord William's attitude. He demanded, in his earliest interviews, the recall of the five nobles, the withdrawal of the illegal proclamation imposing new taxes, the occupation of Palermo by the British army, the command of all military forces in the island, including Sicilian troops, for himself, and the immediate despatch of a Sicilian division to strengthen the British army in Spain. The queen's attitude was equally decided ; all the Ambassador's demands were peremptorily rejected, and he was dismissed the Court with marked ungraciousness. At this moment Queen Caroline was speaking and writing, in apparent sincerity, of the " persecutions" of the English, and the "sacrifices" which the Royal house of Naples and Sicily had been compelled to make to the ambition and greed of England. But against her passionate exclamations we have to set the attitude of the Crown Prince, her eldest son, and the adherence to our cause of the Princess Amélie, and her consort, the impeccably sagacious Louis Philippe.

While the action of Lord William Bentinck was decided, and the action of the queen not less decided, there yet remained the hope that the British Cabinet might interfere to modify the trenchant language employed by the Ambassador. Napoleon was planning another of his innumerable attempts to capture Sicily. If Sir Edward Pellew's fleet, watching Toulon, were scattered, the discontent in the island might provide the most favourable opportunity for the invasion which had as yet arisen. The quarrel, so early and so angrily begun, between the queen and the Ambassador was therefore watched with anxious interest in Sicily and in most other countries of Europe. If the Ambassador were recalled or directed to moderate his language, the Sicilian Question would become acute in a few weeks time.

Lord William had landed in Sicily at the end of July 1811. Within a month he was under sail for England, not having deigned to pay further court to the king and queen since the disagreeable interview in which, to all appearance, the queen had decidedly the best of the encounter. During his absence the work of the Embassy was unprovided for by any special nomination, but the command of the army devolved on General Maitland. Maitland issued an order of the day in which he took the whole army into his confidence. Lord William Bentinck, he announced, had left for London on the most important and critical business, and the commander-in-chief must rely upon the army for its support in whatever crisis might arise during his absence. What this grave statement might portend remained a matter of conjecture until December 7, 1811, when Lord William returned to the Embassy. He proceeded in the deliberate manner of a man who is assured of his position; cleared off arrears of work, attended to numerous points of detail, and did not seek an audience with the queen until January 1812, or nearly a month after he had returned to duty.

The queen, who was fearless as Bentinck himself, granted the interview. The moderation of Lord William's language was as marked as his former abruptness. But that very moderation told Queen Caroline only too clearly that the patience of England was exhausted. If Lord William had been abrupt and peremptory before, it was because negotiations were still possible; or at least he professed to think that they were possible. On this occasion he made no further proposals, he did not even repeat those which he had so forcibly urged six months before. He merely implored the queen, as a friend—if he might so far venture—to leave Palermo, and, if possible, to leave Sicily altogether. The daughter of Maria Theresa struggled to the last. If anybody was to leave Sicily, she retorted, it would be the meddlesome English. Let them begone! She and her Sicilians could defend the island without their help. The distressing interview lasted for three hours.

The queen very likely believed what she was saying. The stress of misfortune, lasting unrelieved for nearly twenty years, had really weakened her judgment. Finding in this conclusion the gentlest explanation of the many painful scenes of this epoch, the historian will probably cease from probing the recesses of that dimmed and saddened mind. Bentinck's course was clearly marked out. From the queen he turned to the king. Profound secrecy was observed as to the nature and course of the negotiations which ensued. The consequences only were marked; but these were of so striking a nature as to leave little doubt in men's minds as to what had passed. The queen withdrew from Palermo, and informed the Austrian Minister that she intended to retire to Vienna when milder weather set in. The king, never of much consequence, now formally made over his powers to his eldest son, the Crown Prince ; a decree of even date conferred the command-in-chief of the Sicilian forces upon Lord William Bentinck, and empowered him to garrison Palermo with British troops in whatever strength he might see fit to order. The illegal impositions were formally abrogated, and the banished nobles recalled to Sicily. The news of the two last-named measures was received with general satisfaction and even rejoicing—somewhat moderated by the reflection that Sicily had found a new master, and that the master was the unknown quantity, Lord William Bentinck.

Probably the master of Sicily was prepared for at least one more struggle before his power was assured. The concession that the queen should leave Palermo was not enough. He was aware that so influential a character could not be other than a centre of disturbance in the island ; and a centre of disturbance Queen Caroline continued to be for another year.

It was in the line of policy laid down after February 1812 that Lord William Bentinck displayed the limitations of his intelligence. Up to that date his action had been unexceptionable ; it had called for nothing but courage and resolution. After February 1812, when the queen was no longer able to exercise direct authority, and when the island was almost denuded of Sicilian troops, it became a question of how Bentinck would use the absolute authority with which he had insisted (quite rightly in the circumstances) that he must be clothed.

England, at the time, commanded the services of an administrator of a high order, perhaps of the highest order. She had employed him as first British Governor of Malta, where he was struggling with a task as difficult as that which faced Lord William in Sicily. Maitland's method of setting to work was, to master the question he proposed to attack down to its most trifling detail : after which he proceeded to action, moving more or less cautiously, according to the complexity of the problem. Bentinck's method was materially different. He would survey the conditions of the problem—men, institutions or measures—decide the relations in which they ought to stand to each other in the future, and then command that they should assume those relations. But the conditions were too complex for this Jove-like attitude. Men as proud as Bentinck himself, a state as old as England itself, could not be made to bend this way and that at the nod of a despot. Incidentally, his irresponsibility led him to over-look some serious irregularities in the administration of the law under English supervision. And yet it was the legal system which ought to have first attracted his attention. For it was by able use (or misuse) of the facilities afforded for legal persecution in Sicily that the Court party contrived to make the life of any adherent to the English cause a burden to him. Thus Bentinck's authority was no sooner established to all outward appearance than it was undermined.

This is but one example of many which might be cited to show, not that Bentinck lacked courage, but that he lacked that nameless quality which helps a man to seize the right moment, the right man, the right measure, and makes him a successful, instead of an unsuccessful administrator. It would appear, moreover, as if Bentinck allowed his mind to reflect only too faithfully the divergent opinions on the question of the British occupation of the island which were current in well-informed (if not authoritative) circles. The only ground on which the British control of Sicily could be justified was that it was imperative, if Napoleon's plans were not to succeed. Although England thereby guaranteed at least one throne for Queen Caroline it would not have been justifiable on that ground only. It may be safely affirmed that from first to last the British Government had no object whatever in incurring the vast expense of defending Sicily, except heading Napoleon off from the East. But many influential men were beginning to grumble at the notion of ultimately evacuating the island, and leaving it to the tender mercies of a sovereign who had received millions from the English, had repaid them with abuse, and had intrigued (as was firmly believed) with the arch-enemy.

Bentinck, himself, leaned strongly to the latter view : and he allowed his prepossessions to dictate his actions. Not that his action was wise from any point of view; for he took in hand the reform of the Sicilian Constitution. This measure, which would have been one of the most complicated nature, even supposing that the island was destined to be British ground, was totally uncalled-for, if England was doing no more than temporarily garrisoning Sicily in its own interest and against the common enemy. Within six months of his return from England, the reform of the constitution was dictated in a speech from the throne delivered by the Crown Prince at the opening of an extraordinary session of the Sicilian Parliament.

The quarrels of the king with the queen, and of the nobles with both ; the question of the command of the army, the pretensions of the Church, the discontent of the populace—any one of which questions, taken separately, would have required some months of careful attention—were thus raised publicly, *en bloc*, and in the manner calculated to give the maximum of offence and the minimum of satisfaction. The Church, its interests safeguarded, looked on with chilly indifference. The nobles were furious at the direct attack (justifiable or not) upon their privileges, and the people were not yet ripe for the radical changes inspired by Lord William Bentinck. As if he had not enough on his hands already, Lord William raised, in its most contentious form, the blazingly contentious question of the relations of Sicily to Naples.

This measure, the reconstruction of the Sicilian Constitution, ruined, for the second time Lord William Bentinck's reputation as a constructive statesman. But this is the smallest part of the damage it effected. It made the government of the island more difficult than ever. Even then we are only at the beginning of the mischief. The worst effect of the attempted reform was that it destroyed all confidence in England. Incidentally it showed up the Crown as the mere puppet of the

Ambassador. There was more in the Crown Prince than was supposed ; but the fact that he could be induced to recommend so wild a scheme in a speech from the throne materially lowered his authority and influence.

The loss of confidence in the English was very serious. They had been compelled to take steps which could not but be odious to the Sicilians—to depose their king, and to banish their queen. In common prudence they should have stopped there ; and should have allowed the islanders time to assure themselves that the intentions of England were conservative. Instead of which a red revolution was inaugurated. In so far as the new constitution was operative it was dangerous : in so far as it was inoperative it was ridiculous. It was remarked at the time that it displayed only two defects —ignorance of Sicily and ignorance of England. It was a most deplorable blunder. It is to be observed that in this year, 1812, England began, through its representative, to undo, as rapidly as possible, the reputation for fair dealing which had justly been claimed during the preceding twenty years. Up to 1812 the detractors of England were obviously in the wrong : after 1812 they came to be more clearly in the right with every month that elapsed. The year 1812 was also the year of the campaign of Moscow, the beginning of the end of the Napoleonic empire. The efforts of England were about to be crowned with success : it is distressing to reflect that but for the events of the two years 1812–1814 unblemished honour, as well as success, might have been claimed for England. While the Crown Prince was recommending the Sicilian Parliament to consider measures which he could not but feel were fraught with the most alarming possibilities for the future of his father's kingdom, the King of Naples was already operating on the Vistula. Thus at the moment of greatest apparent peril for Sicily, the days of her deliverance were drawing near. The Spanish campaign had been one false step ; the Russian campaign was the second and most momentous mistake of the great emperor. It has been customary of late years to maintain that the Russian campaign was really aimed at England, that it was forced upon France by the determination of this country not to submit to the decrees of Berlin and Milan.

It is natural to hesitate before accepting a conclusion which gives England so prominent a place in the great struggle. Nevertheless, so far as the evidence is forthcoming, the conclusion does not seem to be strained; and the fate of the French kingdom of Naples is a striking example, on a small scale, of the effect produced throughout Europe by the naval predominance of England.

For the moment there was no sign that the end was near. In Naples, Queen Caroline acted as Regent for her husband, King Joachim; in Sicily, the Crown Prince acted for his father. Neither kingdom was happy. Sicily was in a state of apprehension approaching a reign of terror, and Naples was bowed down under the weight of the still increasing taxation caused by her participation in the Russian campaign. At first there were good hopes that the war-taxation would cease; for the beginnings of the Russian campaign were fortunate for France, and the King of Naples covered himself with glory.

The Russian campaign, momentous as it was for the fate of Europe, had remarkable consequences in Naples. The king had long been cruelly mortified at the treatment which he had received at the hands of his brother-in-law. He had fought magnificently in Russia, but had frequently disagreed with the emperor during the course of the campaign. He returned to his kingdom, with his mind made up. He would endure no more interference. Coincidently with this resolve he had taken another: he would not allow his kingdom to disappear in the wreck of the French Empire. What the emperor would not consent to see himself was perfectly clear to King Joachim. Napoleon could not hope to face the Continent in arms. Even the resources of France were becoming exhausted under the strain which the sovereign ruthlessly imposed upon them. It was clear that if the emperor would not temporize, the empire would soon be in direst peril. All these reflections pointed in one direction—they pointed to the King of Naples coming to terms with the Allies.

A totally new situation would then be created in Italy. Hitherto it had been war to the knife between England and the French kingdom of Naples. Although England was no longer proud of fighting for King Ferdinand, there was no help for it : for King Ferdinand England must fight, if the only alternative was to be the French kingdom of Naples, with a sovereign who was the puppet of the Emperor Napoleon. But if it were a choice between two kings, one Ferdinand and the other Joachim, the one a worthless idler, the other brave, brilliant and competent, the situation was not so clear. This difficult situation was soon to be presented to the London Cabinet. For any observer not accepting blindly the traditional estimate of the policy of England, it was not hard to predict which of the two monarchs would receive British support. The moral and material effect of detaching the King of Naples from the cause of the emperor would be incalculable : the price—(his recognition as sovereign, with probably an increase of his territory) was considerable, no doubt. But the advantages of securing his support were also very considerable. His price was, in fact, merely a question whether he was indispensable to the Allies or not. Events moved rapidly. The king had been left in chief command of the relics of the Grand Army, after the departure of the emperor for Paris. He had in vain sought permission to return to his kingdom. In the end he handed over the command to Eugène Beauharnais, and without waiting for orders from the emperor, he set out for Naples, which he entered on February 5, 1813. He was welcomed with tumultuous joy. The Neapolitans were in the habit of according a hearty welcome to any monarch who entered their city in due state. But there was more than this in the attitude of the popular mind towards King Joachim. Royal processions were, indeed, no new things for the Neapolitan ; but a king who had achieved everything was a novelty to them. For fifty years, since the departure of Charles the Third for the throne of Spain, there had not been so worthy a figure on the throne as King Joachim. The entire failure to recognize this was only one of many blunders in our Sicilian policy. When we read, throughout Lord William Bentinck's despatches, of the hatred of the Neapolitans for King Joachim, it becomes only too clear that England was represented by a man who substituted prepossession for observation ; and who acted upon his own prepossessions, instead of upon even a casual observation of the facts of life.

The king's popularity found a foil in the decided unpopularity of his wife, who had acted as Regent during his absence at the war. It was enhanced by his marked anti-Gallic attitude. The dependence of Naples upon France had been the only consideration which had restrained many Neapolitans from giving their hearty support to the king. Joachim made it quite clear, by action and speech, that his days of servitude were over. Quickly his own resolve took shape ; and as it took shape it was presented to the Court and the public, with the, perhaps indiscreet, frankness that marked his career. His days of servitude were over ; he would no longer defer to Paris. He would reign as a Neapolitan king—as an Italian king, perhaps. For Naples was still the most considerable state in Italy. Why should not Naples become the head of the united peninsula, and receive, instead of rendering, homage ? Such a programme could not but be flattering to the Neapolitans : in the disordered state of Italy it is impossible to deny that it was feasible (granted some external support) and decidedly to the advantage of the other states. " Some external support " meant, primarily, the support of England. Joachim had watched the resources of his kingdom wasting away under the relentless pressure of British hostility. He was a sufficiently good financier to realize that his resources would be more than doubled if his coast-line were free. The commerce of Naples had been considerable in the days before it was shackled by the Milan Decrees ; it needed only the removal of British cruisers to recover its vitality. After England only one other Power remained to be con-

sidered—Austria. The restoration of some, at least, of the
Austrian possessions on the mainland of Italy would
probably be indispensable if the support of Austria was to
be secured. But leaving the details of his remarkable scheme
to be worked out as events might develop around him, the
king addressed himself with much assiduity and ability to
winning support, for the unification of Italy under his own
guidance. The queen represented, perhaps too ardently, her
brother's interests. She was not very heartily with the king,
in spite of the aggrandizement of her House, which would
have followed on the transformation of the King of Naples
into the King of Italy.

These designs were destined to be baffled by the action of
Lord William Bentinck. But for the moment it seemed as if
Sicily was the last point from which Naples need anticipate
difficulties. The new constitution of the island was formally
approved in London, and Lord William received every
support from the Foreign Office. There is no possibility of
doubting that the scheme was well-meant, both in London
and in Sicily. There were no sinister designs behind the
amazing screen of absurdities to which we had committed the
Sicilian people. But inasmuch as the constitution was the
outcome of ignorance, and was, in point of fact, unsuited to
the situation, it could have but one effect—the paralysis of
the Government. Lord William might feast and perorate
almost daily—for he was an accomplished and magnificent
viceroy—but the fact remained, that the island awaited but
the opportunity to thrust England forth, and be rid for ever
of the English and all their doings.

In the meantime, King Joachim had despatched Prince
Cariati to Vienna. The prince was directed to sound Prince
Metternich on the question of securing the "independence"
of Naples. "Independence" meant the separation of Naples
from the French Empire. Prince Cariati had no written
instructions ; it was clear that the King of Naples had every
intention of moving cautiously. But Metternich welcomed
him with open arms. There was a word said about the

interests of the Austrian Royal family in Sicily; and Cariati went more than halfway to meet the Austrian Minister. King Joachim, he said, was prepared to renounce his claims to be king " of the Two Sicilies." King Ferdinand might certainly continue to be King of Sicily; such a settlement would not interfere with his own ambition to remain King of Naples. The somewhat vaguer stipulation that the King of Naples must fall in with the policy of the Emperor Francis was warmly acceded to. In two interviews with Prince Cariati, Metternich discovered that Austria might count on a powerful, and even an enthusiastic ally.

It was the nature of the King of Naples to be sanguine. But even a man of much less hopeful temper might have concluded that his work was more than half done when he had secured the support of Austria. It would have been natural if Austria had declined altogether to negotiate with him. The Royal family personally had suffered much at the hands of France, from the murder of Marie Antoinette to the beggary of her sister, Maria Caroline. When, therefore, Prince Cariati reported that he had been welcomed with open arms in Vienna, it was reasonable to conclude that the difficulties with England would be at least as easily smoothed over. It was true that England had never recognized Napoleon as emperor; and that she was still less likely to recognize his brother-in-law as king. But, on the other hand, the English Government had maintained only one principle of action in Sicily for twenty-five years; that principle was not necessarily antagonistic to the recognition of King Joachim. The principle was the merely defensive one, that Sicily must on no account fall under the control of the Emperor Napoleon. If King Joachim's proposals were accepted, there could be no danger of this happening; for Sicily would remain a Bourbon kingdom, and Naples would be in the hands of a bitter and irreconcilable enemy of the emperor. As a guarantee of good faith the king further offered to surrender his whole navy into the hands of England.

It was hardly to be expected that England would accede with any grace to the king's proposals. But the spirit of the British Cabinet's reply could easily be anticipated ; it would be that England had no concern with the internal affairs of the Italian peninsula; that Austria was far more intimately concerned in the territorial question, which must necessarily arise in the ensuing negotiations, than England; that England had never had any designs of territorial aggrandizement in the direction of Italy; that provided sufficient security were given that the conquest of Sicily would not be attempted, England could not pretend to veto the choice of a monarch by the Italian people ; and, finally, that, without engaging to recognize King Joachim, England was prepared to cease making war upon him.

This forecast, far from being over-sanguine, erred, if at all, on the side of attributing to England a too lively interest in Neapolitan affairs. The early interest of England in the Sicilian Royal house had been, to a great extent, sentimental ; it had been a chivalrous pride in defending the sister of Marie Antoinette. In the days of Nelson, King Ferdinand was still a young man ; much had been pardoned to him. In 1813 he was an elderly gentleman of sixty-four, and he had alienated all English friends by his heartless frivolity, and by a private life which consorted ill with his advanced years. Misfortune had had the worst possible effect upon the character of the queen. Expressed public opinion in England in 1813 was therefore divided between the views of those who held that any measure would be welcome which would rid the country of responsibilities so irksome, and the views of those who held that the Sicilian Question ought to be solved by annexing the island without further delay. In neither case could there be any reason for opposing King Joachim, who had surrendered—or offered to surrender—all his claims to the Sicilian throne.

A negotiation of such magnitude proceeded slowly. Bentinck received the proposals of the Duke of Campochiaro with reserve, and referred to London for instructions. There was little to find fault with in this attitude; but it was disappointing. As regards Vienna, June 1813 arrived and found no written undertaking as yet drawn up. But the Emperor Napoleon, on whom his brother-in-law's remonstrances had produced no effect, made a fresh demand on Naples—this time for an army corps to replace the losses of Bautzen. The unreasonableness of the demand, not less than the tone in which it was made, not only strengthened the resolution of the king, but cost the emperor his strongest ally in Naples. Queen Caroline might be a good sister; but she had no mind to see her brilliant husband ordered about like a school-boy. Nor could even the Neapolitan Ministers, who still favoured the French alliance, deny that an alliance with Napoleon meant nothing less than servitude.

By July 1813, Lord Aberdeen found himself face to face with Prince Cariati in Prague. The Cabinet had decided to accept King Joachim's proposals, but on the clear understanding that the lead was taken by Austria. These negotiations were interrupted by a fresh outbreak in the interminable war, before Naples had been definitely secured as an ally. King Joachim, who had the faults as well as the advantages of the impulsive temper, inclined once more to the cause of his brother-in-law; but decided on the dangerous course of keeping on good terms with both sides, and looked anxiously for the return of Lord William Bentinck, who was campaigning in Spain. Lord William had found more than his match in Marshal Suchet, and returned to Sicily in September 1813. He was not disgraced—there have been powerful apologists for his campaign—but he had not succeeded.

His most considerable engagement was the combat of the Pass of Ordal, fought in September 1813.

It would appear that some part of Bentinck's failure is to be attributed to lack of foresight, and not to lack of martial ardour. The Marquis of Wellington wrote, on July 1, 1813, in reply to an indent for " spherical case "—" I beg to know whether that article has been left behind as well as everything

x

else!" He thought it well to recommend some elementary precautions to Lord William Bentinck, and, in reply to the latter's report on the alarming condition of Sicilian affairs, the marquis wrote as follows :—

"*Huarte, July* 1, 1813.

"MY LORD,—In answer to your lordship's despatch of June 20, No. 3, I have to observe that I conceive that the island of Sicily is at present in no danger whatever."

Lord William's despatch No. 3 contained the phrase, " Murat has opened up negotiations with us, the object of which is friendship with us and hostility to Bonaparte." Six weeks earlier (May 16, 1813), Lord Wellington had laid before Lord Bathurst his views on the subject of a British expedition to Italy, and had concluded, " Not only should we have no assistance from the country, but as far as their resources would go, the people of the country would assist the enemy."

When King Joachim was once more face to face with Lord William Bentinck the situation was less simple than it had been when the king had approached the Ambassador in secret before the Spanish campaign. The king's view was, that his offer to the Allies had been genuine, but that they had not seen their way to accepting it definitely. It was not to be expected that he should be pledged to them while they remained unpledged to him. After all, the one obligation before him—until he had contracted new obligations—was his duty to the French Emperor ; and his assiduity in fulfilling his duties ought to be a guarantee that the Allies would find him a trustworthy friend if he threw in his lot with them. Lord William's view was precisely the contrary to this. It was maintained that it was idle to talk of negotiations with a man who was one month offering his services to Austria, and the next month was charging Schwartzenberg at the head of the cavalry of the Grand Army. Meanwhile the draft of the Anglo-Austrian-Neapolitan alliance remained no more than a draft. King Joachim was once more anxious for its completion and signature, but the Allies were now much better placed than they had been before the battle of Leipzig. Napoleon was visibly weaker; the Allied armies were drawing nearer to the Rhine.

The situation was serious for the Emperor Napoleon, who no longer transmitted orders or requisitions to Naples. The emissaries demanding reinforcements were succeeded by messengers bringing offers. It was too late. Joachim was determined to preserve his kingdom, and was more convinced than ever that an alliance with France would not help him to that end. Nor were the offers which Napoleon could make considerable in themselves. The trifling increase of territory implied in the addition of Fermo and Ancona to the dominions of Naples were nothing to a man who already saw himself King of Italy. Even Fermo and Ancona were only to be granted if the king would assume the command-in-chief of the army destined to operate on the left flank of the Allies. The command-in-chief would have placed King Joachim over the head of Eugène Beauharnais, and would have left him independent of the emperor himself. But the King of Naples thought first of Naples, and if he were to operate on the left flank of the Allied army, it would be as Prince Eugène's opponent and not as his superior officer. For only from the Allies could he hope to gain that recognition of himself as an independent sovereign which would leave the kingdom of Naples standing when the French Empire had disappeared.

The unfailing touchstone of every Neapolitan complication was the attitude of the Duke de Gallo. In every difficult situation the Duke de Gallo had invariably been found to be on the winning side. It was, therefore, encouraging for the king to observe that Gallo was working as hard as possible for him, and was sanguine of success. Full powers were granted to Prince Cariati to re-open negotiations on the basis of the informal exchange of views which had taken place at Prague between the prince and Lord Aberdeen in the preceding July. This was as much as the king could do for the moment. The initiative rested with Metternich, who might, or might not, be disposed to pardon the king's last campaign against the Allies. Without pressure the assent of Austria would have been difficult to obtain ; and the King of Naples could apply no pressure. Marshal Bellegarde the Austrian commander-in-chief in Northern Italy, came to the king's help by refusing to make a move forward, unless he were definitely assured of the active support of the Neapolitan forces. Bellegarde had replaced General Hiller, who had been removed on account of some unsuccessful manœuvres in the Alps which he had directed his army to make. Consequently, unless the Cabinet of Vienna was prepared to make another change in the command-in-chief, it must needs follow the advice of Marshal Bellegarde.

In the direction of England the King of Naples had been able to effect but little. He had opened communications with Lord William Bentinck, who had flatly declined to continue them, adding that the British Government would never allow the throne of Naples to pass away from the Bourbon family. This was an unpromising temper; but the king reflected that, after all, it was with Lord Aberdeen, and not with Lord William Bentinck, that the last word rested. His offer had been to conclude either an armistice, or a definite peace, or an alliance with England ; and he had empowered his emissary to conclude whichever of these three might fall in with Lord William's instructions. Very considerable commercial concessions were offered as the consideration for the king's recognition by England. It is important to notice the date of this negotiation—November 1813. Lord William, as we have seen, repulsed every approach, asserting the unalterable determination of England to preserve the throne of Naples for King Ferdinand. The negotiation was closed on December 30, 1813, by the return of King Joachim's emissary to Naples.

On December 3, 1813, Lord William Bentinck indited the despatch which was destined to colour the history of English dealings with Sicily. It had a retrospective effect, and not unjustly imparted a sinister character to the dealings of

England with Sicily. It is one of the most remarkable letters ever signed by a responsible official. Without instructions from his Government, and in direct opposition to what (he was shortly to learn) were their intentions, he proposed to the Crown Prince Regent that Sicily should be ceded to England.

So adventurous a proposal, if made in conversation, and after dinner, might pass for an indiscretion—à considerable indiscretion. But it was made in writing, and enforced by arguments, such as the poverty of the island, the difficulty of securing an adequate return for the expense of governing it, and the manifest success of England in administering its affairs. The offer went into details; specific sums of money were mentioned, or, as an alternative, the states of the Church would be considered by England to be a fair equivalent. They might be added to Naples if Sicily became an English island.

The Crown Prince took the communication seriously—he could hardly do otherwise—and begged for some reference to the Ambassador's instructions bearing on the question.

Lord William replied that he had no instructions on the subject at all; and gathering, perhaps, from the short and guarded reply of the Crown Prince that the proposal was not entirely to His Royal Highness' taste, he was not sparing of words. The proposal, he added, was "a phantom of his own disordered brain;" it was a "sogno filosofico," "le rêve d'un voyageur."

The Crown Prince Regent of the kingdom of the Two Sicilies is not to be blamed for disbelieving Lord William's disclaimer. The Ambassador was in command of a very large armed force in occupation of the island. He had so managed affairs that he was practically dictator; the Crown Prince had good reason to know that the word of Lord William carried greater weight than the word of the nominal sovereign. For many years past England had been actively concerned in the defence of the island; of late years all Sicilian institutions had been re-cast on English lines. It was inconceivable that a functionary so highly placed should speak so lightly of territorial changes of such magnitude. Nor was it unreasonable to suspect that England was now preparing to secure the reward for her services, and that the acquisition of Sicily might be that reward. England had already taken Malta; the Ionian Islands were falling, one by one, into her hands. Queen Caroline had always maintained that there was nothing to choose between France and England : France wanted Naples, and England wanted Sicily. There had recently been much talk in England about the desirability of making Sicily English ground. The Crown Prince was not to be blamed for concluding that all these considerations told in the same direction ; and whether he is to be blamed or not, he took the practical step of directing Prince Castelcicala, who represented King Ferdinand's Government in London, to demand a definite statement from the British Foreign Office. Prince Castelcicala obeyed his orders. Commenting on the phrase " le rêve d'un voyageur," he added that an Ambassador who was subject to dreams of this kind was not a fitting person to be accredited as the representative of Great Britain. In the circumstances he could not do less, he added, than demand Lord William's immediate recall. The proposal by an Ambassador to acquire a country to which he was accredited, being without precedent, he could do no less than take this serious step. It was not sufficient reparation for the Ambassador to say that his proposal was a philosophical dream.

Prince Castelcicala's request for the recall of Lord William Bentinck was not acceded to. But the official disclaimer of the Foreign Office was prompt and decided, and was accepted by the Crown Prince Regent.

The proposal of Lord William Bentinck to acquire Sicily for England by exchange or purchase was disastrous, not only because it was, in itself, an absurdity, but because it gave a questionable colour to British conduct—past, as well as future. Official disclaimers of individuals are, or were, not uncommon incidents of public life. On this occasion nobody concerned could avoid the conclusion that the proposal, and not the disclaimer, really represented the policy of the British

Cabinet. "This, then, was the true objective of the last twenty-five years of English activity in the Mediterranean. This, therefore, must necessarily continue to be their objective in the future"—so must have reasoned any Sicilian who was in possession of the facts. Yet, so far as the evidence goes, these conclusions were entirely faulty; the disclaimer was genuine, the proposal unauthorized and disapproved. The action of England in the Mediterranean from the outbreak of the French Revolution down to the year 1815, was not dictated by European, still less by dynastic, considerations. This was shortly to be proved in the most unmistakable manner by the recognition of Joachim Murat as King of Naples. The inspiration of all English efforts, naval and military, in the Mediterranean during these years, was the paramount need of keeping the way open to the East; or, at least, of preventing the way to the East falling under the control of France. In this large policy it was really a matter of no concern to England who reigned in Southern Europe, so long as the influence of Napoleon was excluded. England had no desire for more territory than sufficed for the command of the waterway.

While Lord William Bentinck was assuring King Joachim's envoy that the British Government would never consent to recognize Joachim Murat as King of Naples—while Lord William was sounding the Crown Prince of the Two Sicilies as to the surrender of Sicily to England, when the Royal family should once more have taken up its abode in Naples, very different views were being canvassed and accepted in London. The Emperor of Austria, giving effect to the representations of his commander-in-chief, had decided to recognize King Joachim with the object of gaining for Marshal Bellegarde the support of the Neapolitan army. England had decided that military considerations were paramount, at least until the Allies had entered France, and was now prepared to accept the lead of Austria in Italy. Instructions were therefore despatched to Lord William Bentinck, directing him to conclude a peace with Naples, and to recognize Joachim

Murat as king. This despatch was supplemented, a month later, by a second despatch, in which Lord Castlereagh went into particulars. Naples, said Lord Castlereagh, must be considered to have departed for ever from the Bourbon House of the Two Sicilies. Henceforth Ferdinand could be King of Sicily only. But although the logic of facts compelled the Powers to take up this harsh attitude, there was no intention of despoiling the Royal house. The Crown Prince Regent was invited to choose which, in order of preference, he would select from the following "compensations": Poland, Lombardy, Saxony, Sardinia, Corsica, or the Ionian Islands. One more choice was offered to Sicily, perhaps the most remarkable of this extraordinary list. The Crown Prince was assured that he might compensate himself for the loss of Naples by making a selection from the West Indian Islands. Few incidents of this epoch display more graphically the break-up of ancient ties, and the solution of ancient obligations, which had taken place since the outbreak of the French Revolution.

In order, therefore, to secure the adhesion of the King of Naples to the cause of the Allied armies, his title was to be recognized by Austria and England—practically by Europe; for, in 1814, where Austria and England led, the rest of Europe must needs follow, especially when Austria had undertaken to secure the recognition of King Joachim by the Tsar. King Ferdinand was to remain in Sicily; this would be enforced by the armed interference of Austria, if necessary. When this decision was intimated to Lord William Bentinck in January 1814, he might naturally have reflected that his work would already have been half done, if he had not been so peremptory in his rejection of King Joachim's advances in the preceding November. His high language was not only out of place, but was in opposition to the intentions of the Government—intentions which were now embodied in definite instructions.

In reviewing his own career in Sicily, Lord William could hardly have avoided some chastening reflections. His Spanish campaign had been a failure, though he had been in

markedly superior force to the Duke of Abufera. The Duke
of Wellington had not attempted to conceal his small opinion
of Lord William's military capacity. His unauthorized offer
to the Crown Prince was in direct contradiction to the tenor
of the instructions which reached him in January 1814, and
he could not but expect that it would damage him with his
own Government and with the Government of Sicily. He had
originally been despatched to Palermo in order to see that the
British subsidies were not wasted, and that British authority
was not undermined by those whom it was designed to
protect. That much he had effected—and so successfully
effected that Queen Caroline left Sicily for Vienna in June
1814. The king had been compelled to abdicate—no great
sacrifice on his part—and had consoled himself for the loss of
the queen's society by enjoying that of her destined successor,
the lady who was afterwards created Princess of Partanna.
The Crown Prince was apparently docile—although less
docile than he appeared. It was not in action that called for
courage, and courage only, that Lord William fell short; it
was in action that could only be taken after a certain amount
of intelligent observation. And when we are asked to believe
that Lord William was merely self-deceived when he assured
the Crown Prince of the excellent results of British administration, it is only too easy to believe that such was the case.
The man who could force a constitution on an unwilling
people was the very man to maintain that it worked well, in
the face of its deplorable effects. So many constitutions have
lived their short life since 1789, and disappeared in revolution
and bloodshed, that their history is familiar to us all.
Deputies elected by a fraction of the electorate squander
public money and promote private ends, while the sullen tax-
payers look on, suppressing their rage till the convenient
moment for taking action arrives. In the meantime secret
societies flourish ; there are riots and lynchings. If there is a
scarcity there are premature revolutions, officially described as
local discontent and disorder. Such was the mockery of
"popular government" in Sicily, which Bentinck was proud to

have initiated, and to whose excellent results he pointed when he proposed to make the British occupation permanent. Bentinck, therefore, had every reason to welcome the instructions which reached him in January 1814. They relieved him of a responsibility which he may have enjoyed, but for which he was patently unfitted. They supplied a simple solution of a most complicated situation. Of Bentinck's own efforts to meet that situation, the best that can be said is that the sooner they were forgotten the better. And if there were no other reason for obeying his orders, common honesty dictated that he should do so, unless he was prepared to resign his post, and the income of £14,000 a year which he was paid for representing the British Government.

Lord William Bentinck, however, decided to disobey orders; he reproved the British Cabinet for the treaty with Austria, and prepared to postpone by every means in his power the recognition of Joachim as King of Naples. His own expressed desire was to restore King Ferdinand to Naples, and keep Sicily for England. At the same time he was writing to Lord Castlereagh that he was weary of the work, worn out with the difficulties of the situation, and anxious only to resume his military career. If this were true, a very easy path lay before him; to obey orders would cost him the minimum of trouble and anxiety. As he put himself to vast inconvenience, and seriously embarrassed the Cabinet by refusing to recognize King Joachim, the simple conclusion is that he was deliberately playing a double game. More charitably it may be said that he was a man full of undisciplined, untaught energy; incapable of observation, impatient and dictatorial, and expressing with equal violence the most varying opinions, careless—with the carelessness which was part of his arrogance—what other people might think of him; and continually vaunting his own "honesty," "frankness" and "straightforwardness."

Those negotiations now opened which fill so lamentable a page in the history of British diplomacy. In furtherance of the treaty between Austria and England, arranged by Prince

Metternich and Lord Aberdeen, there arrived at Naples, Adam, Count Neipperg. The count was Ambassador Extraordinary, charged, on the part of the Emperor of Austria, with the duties of recognizing Joachim Murat as King of Naples, and arranging, in concert with the king, the campaign of the Allies in Northern Italy. Count Neipperg was merely an instrument, and there is no apparent ground for Lord William's charge against him of having been " overreached " by Murat. Nor is it possible to understand how Lord William could reconcile it with common honesty, to say nothing of courtesy, to describe the king as a " criminal "—" whose whole life has been crime," is the exact phrase.

The King of Naples despatched an aide-de-camp, Colonel Barthémy, and Baron d'Aspern, of Count Neipperg's suite, to Palermo. These representatives, the one of Austria, the other of Naples, were empowered to carry out the formal part of the treaty. The English Ambassador declined to have any dealings with them, on the ground that he would be compromising himself. But he despatched his private secretary, Mr. Graham, to Naples, charging him to find his way somehow to the head-quarters of the Allied army. Mr. Graham was to secure a passport by agreeing to an armistice ; and being arrived at Geneva, was to inform the Allies that Lord William Bentinck was about to invade Corsica with a force of 10,000 foot, 400 horse and 30 guns. The authority for this undertaking may be found in Lord William's commission as commander-in-chief of the naval and military forces in the Mediterranean, exclusive of Malta. But the immediate duty before him was the recognition of King Joachim. Having refused to treat of this with Colonel Barthémy and Baron d'Aspern, it was not likely that he would grant extensive powers to Mr. Graham. Mr. Graham, in fact, had no powers ; and a series of irritating interviews took place between him and Count Neipperg. These interviews were the beginning of that profound mistrust of England which culminated in Count Mier's outburst that he did not see how England could expect any progress to be made, unless the Government would employ men who would pay some attention to their instructions.

The king spoke out plainly. His navy was England's, he said (by the mouth of the Duke of Campochiaro), whenever England chose to take it over. In alliance with England he would maintain order in the Italian peninsula. Naples had no desire for any other ally. Mr. Graham would say nothing, could say nothing.

On January 7, 1814, the Duke of Gallo and Count Neipperg concluded the business, so far as Austria was concerned. There were secret articles in the treaty signed on that day. In accordance with the line suggested by Lord Aberdeen, Austria appeared as the Power taking the initiative. A secret article bound Austria to obtain the recognition of King Joachim by England. Another secret article pledged Austria to obtain the resignation of King Ferdinand's claims on Naples. This article Austria was prepared to enforce by arms. In watching the antics of Lord William Bentinck, we must never lose sight of the clear and intelligible line taken up by the English Cabinet on this question. In accordance with the line of conduct pursued throughout by England, it was made perfectly clear that the question was not, primarily, one that concerned England at all. It was the throne of an Austrian princess that was at stake, not the throne of an English princess. Austria had territorial interests in Italy : England had none. If Austria were satisfied that the presence of an enlarged South Italian kingdom would prove to be a guarantee of order, it was not for England to withstand her policy. England made no pretence to a leading part in the affairs of the Italian peninsula, and Lord William was not called upon to criticise Austrian policy ; still less was he entitled to complain of Prince Metternich's "apparent want of good faith." The treaty signed, Count Neipperg proceeded to recall Lord William to a sense of his duty ; reminded him of the instructions of his Government, of the disordered condition of Italy, and the critical state of European affairs ; and pointed out that, for the moment, the responsibility lay with him, Count

Neipperg, and Lord William Bentinck. Would not Lord William sign?

Lord William Bentinck would not sign; but he consented to betake himself to Naples, provided he was received incognito. On February 6, 1814, he made his way thither and interviewed the Duke de Gallo and Count Neipperg personally. He reported to Lord Castlereagh that those personages were anxious for him to sign (not unnaturally, seeing that it was his duty and theirs), but that as he was convinced that no reliance could be placed upon Murat, he had declined to do so. He returned to Palermo.

Lord Aberdeen had anticipated some delay in settling the treaty in so far as Austria was concerned. The possible points of contention were numerous and important. Perhaps for that very reason he had urged that Austria ought to take the lead, so that when once these difficulties were settled, the adhesion of England—the unconcerned power—might follow automatically. He was greatly relieved to learn that the treaty was signed with Austria; he called it "the best intelligence," and concluded that the object of the Allies was gained, and that Bellegarde, assured of the support of King Joachim on his left flank, would at last be able to set his army in motion against Eugène Beauharnais. But Bentinck's menacing attitude undid all the good of the treaty with Austria. We know now that Bentinck was actually meditating his extraordinary descent on Genoa; but, at the time, his considerable preparations appeared to be directed against Naples. The violent language that he allowed himself to use against King Joachim made it only too probable that he was contemplating a descent on Naples. In these circumstances it was impossible for the king to denude his country of troops; and Bellegarde's advance was, in consequence, postponed. Thus Bentinck managed to nullify the efforts of Austria, as well as to evade the duties laid upon himself. With Bentinck refusing to recognize King Joachim, openly inveighing against the duration of the Bonaparte kingdom, and assembling a large army with unknown views of conquest, it is not wonderful that the king should ask himself what was the meaning of the clause in the treaty stipulating that Austria should secure the assent of England? Had secret instructions in a contrary sense been sent to Bentinck? Was the treaty a trick to draw his army away from Naples while Bentinck invaded his country? His representations met with the only possible answer from the Austrians—that they had done everything in their power, and could do no more. They quite admitted that there could be no campaign in the north of Italy, until it was certain that the south was secured by the neutrality of England.

In the meantime the king was allowed to assure himself of the promised increase in his territory. On January 19, 1814, the troops of King Joachim had entered Rome. The lordship of Ancona was secured at the same time, amid general satisfaction. The French troops of occupation were withdrawn ; a visible stride had been made towards the ideal of a strong Italian kingdom. Even Bentinck could not deny the existence of a current of popular feeling in favour of King Joachim. " All parties," he wrote, " agree in one view, viz. that of augmenting as much as possible Murat's power, and of uniting Italy under his standard." What " all parties " are " agreed " on, is the best government for a country according to any theory of government favoured by Lord William Bentinck. He ought, therefore, to have been thankful that his instructions and his theories were alike agreeable to the state of things with which he had to deal. But no. " A stand should at once be made against these views of ambition "—such was his comment in March 1814, two months after he had received the direct command to recognize and support King Joachim.

The unification of Italy designed by Austria and England (afterwards so rudely opposed on the same question) was defeated by Lord William Bentinck ; but not unconsciously. We have seen that " all parties " were rallying round Murat in March 1814. Even earlier Lord William had written, January 1814, " When the viceroy is driven back to the Alps the Italians will certainly gravitate towards Murat." Precisely :

it was upon the attractiveness of the king's personality, and the success of his government, that Austria and England relied to decide the "gravitation" of the Italian population. "But," continued our Ambassador, "if the British protection and assistance had happened to be within their reach, that great floating force would certainly have ranged under their standard. The national energy would then have been roused, like Spain and Germany, in favour of national independence, and this great people, instead of being the instrument of the ambitions of one military tyrant or another, or, as formerly, the despicable slaves of a set of miserable petty princes, they would have become a powerful barrier both against Austria and France, and the peace and happiness of the world would receive a great additional security—but I fear the hour is gone by." On the contrary, the hour had come ; and the man, by Bentinck's own showing. If the somewhat incoherent despatch just cited has a definite meaning, it would imply that Bentinck supposed that Italy could be unified under England's leadership, and in hostility to France and Austria. Surely a wilder dream, a tissue of blanker impossibilities was never penned by a serious statesman. Unified, Italy might certainly have been in 1814 but for Lord William, but only upon the lines agreed upon by Austria and England. Why Lord William Bentinck played a part so actively hostile to the ideal he professed to believe in, has been variously explained. The obvious explanation was that Lord William would have been pleased to lead the unification movement himself, but could not endure to act a secondary part. Another explanation is that he intended to be Viceroy of Sicily, however much his ambition might cost Italy. But the statesmen with whom he dealt did not trouble to penetrate so far into Lord William's mind. They rested in the conclusion that if he was in his right senses he was patently dishonest. Bellegarde inclined to the view that he was not responsible : Mier to the view that he was responsible, but was not acting straightforwardly.

In the meantime, Naples had broken definitely with the Emperor Napoleon. The occupation of Rome and Ancona was a measure too grave to be overlooked. On January 22, 1814, the French Ambassador asked for and received his passports; Naples was at war with France. The king moved north in command of his army; and a general exodus of French officials testified (to the contentment of Neapolitans) to the unalterable resolution of Joachim to reign henceforth as an Italian king.

Lord William, on his return to Sicily, still under the delusion that all was for the best there, wrote definitely to Lord Castlereagh that at all costs Sicily must remain English. This, not because the island in French hands would be a menace to English interests, but so that Sicily might remain as an example of how a country ought to be governed. Vanity and self-deception can go no further.

The alliance of Austria with Naples had not been ratified. To have formally completed the alliance in the short space of time which was available, if the troops of Naples were to be of service to the Allies, would, in any case, have been an effort for the Austrian Foreign Office. The obstacles thrown in the way by the English Ambassador formed an additional and unexpected source of delay. Nevertheless the king had, on the whole, pledged himself to the Allies; although there were murmurs that he was not energetic enough. No such complaint was made of Queen Caroline who was left behind in Naples as Regent. Whatever steps could be taken to make the breach with France definite, were taken by her promptly and energetically. She expelled all French subjects: annexed Ponte Corvo (the principality of Bernadotte) and Benevento (the principality of Talleyrand) to Naples, and seized all French vessels in Neapolitan waters. In the capital the Regent ruled with energy, sagacity and promptness. In the field the Austrians and Neapolitans mutually reproached each other. The king stated that Austria should have shown him more consideration. The painful resolve to break with all the traditions of his life, and to make war upon the emperor he had so often followed to victory, ought not to have been made still more painful by leaving him in doubt as to whether he had gained new friends by his dangerous resolve. The Austrians, expecting to find the cavalry-general they knew so well, found only an anxious king poring over treaties, and holding night-long conferences with envoys. Bentinck decided that each was trying to overreach the other, and that Bellegarde was as bad as either of them. "In point of fair dealing I consider Prince Metternich and King Murat to be nearly on a level." "I found Marshal (Bellegarde) anxious to believe to be true that which he knew to be false."

Marshal Bellegarde had nothing to gain from Lord William Bentinck, so there was no reason why he should allow himself to be hectored. Accordingly, when Lord William approached him in the temper of suspicion and contempt implied in the phrase just quoted, the marshal cut Lord William short with a plain reference to his instructions. He followed up the reference by a reminder that it was Bentinck's duty to keep on good terms with the King of Naples. From Bellegarde Bentinck turned once more to Gallo; to whom he delivered, by the hand of Sir Robert Wilson, a communication so offensive in its language and tone that the king declined to hold any further communication with the British Ambassador. For the future, Gallo was commanded to say, the Neapolitan Government would communicate directly with the British Cabinet.

This incident brings us to April 2, 1814, on which date Lord William wrote to Lord Castlereagh, commenting on the conclusion of his dealings with the Court of Naples, as follows :—" I have resolved to be no party to a system of weak and timid policy, which, in my judgment, promises no material present advantage, and certainly none to counterbalance the dangerous effects of Murat's ambition." This extract contains in three lines so many and so serious misconceptions as to Lord William's position and duties, that it must be allowed to speak for itself. A phrase which saves comment on Lord William's behaviour is the more welcome, because there remains much to which some allusion must necessarily be made. It is impossible to pass unnoticed Lord William's elaborate insolence to the king—insolence of manner and

Y

speech. It is impossible not to allude to Bentinck's threat
to restore King Ferdinand, if his own conditions were not
complied with : a threat in direct violation of his instructions.
These are but two of many instances of the deplorable bearing
of the representative of England. It was little to be wondered
at if Murat thought himself betrayed. The British Cabinet
said one thing ; their representative said precisely the opposite.
The king was informed that a certain course of action had
been laid down by Lord Aberdeen and Lord Castlereagh. But
their agent would have nothing to do with Lord Aberdeen
and Lord Castlereagh. Which of these two courses represented
the real intentions of England ? It is not to be wondered at
if the king, finding himself thrust back by the Allies, allowed
Eugène Beauharnais to approach him : at first in secret,
afterwards more openly.

Meanwhile Lord William, as commander-in-chief of the
British forces, decided to carry on a campaign on his own
account. He detached Colonel Montresor to seize Corsica,
and embarked himself for Genoa—which city capitulated to
him on April 18, 1814. The restoration of a state of things
in that ancient city, made unauthorizedly by Lord William
Bentinck, in the name of England, was a serious source of
embarrassment to the great Powers when Genoa came under
consideration. But, for the moment, the chief reason for
noting the campaign of Genoa is the behaviour of King
Joachim on the occasion, and the attitude of Bentinck in
return.

The capture of Genoa was not a very considerable military
achievement; but King Joachim decided so to regard it.
Undoubtedly it placed Lord William Bentinck in the position
of leading the van of the invasion of France from Italy ;
although, as Genoa did not fall until April 18, 1814, a clear
week after the abdication of the emperor, its capture had no
important strategic results. In the capture of Genoa and
Corsica, King Joachim regarded the dramatic rather than the
political effect. To bring once more under English rule the
island which Elliot had been forced to abandon twenty years

before, in the early days of the Revolution, to enter in triumph a place of the historic renown of Genoa—these were the events on which he chose to congratulate Bentinck. The king's letter of congratulation was written, as he explained, from one soldier to another. As a king, His Majesty continued, it would be impossible for him to overlook the language addressed to him; but as a soldier he could not withhold his tribute of admiration for a fine exploit. Together with his congratulations, His Majesty begged Lord William's acceptance of a sword. As there was not time to have such a sword prepared as would be handsome enough for the occasion, the king begged Lord William to accept his own sword.

To be complimented as a soldier by the hero of countless fights was, in itself, a distinction for a commander of Lord William's mediocre abilities. The sword of Murat was a present which kings might have coveted; while the manner in which it was offered was grandly courteous. It is odious to be compelled to record Lord William's reply. He accepted, having been recently rebuked by Lord Castlereagh for his improper behaviour; but he accepted with bare civility, and in narrating the incident to Lord Castlereagh he wrote :—" It is a severe violence to my feelings to incur any degree of obligation to an individual whom I so entirely despise."

There is something great in arrogance when it towers to these heights ; but so large a measure of any virtue is apt to be inconvenient in public life. Lord Castlereagh found it to be so in Lord William's case. The Ambassador concluded his despatch by saying that perhaps the Prince Regent might like to keep Murat's sword as a curiosity.

If there be one single regret mingled with the delight with which the civilized world regards Unified Italy, it would be that this illustrious contribution to the happiness of the human race could not be made without the sacrifice of Murat. We have too long been acquainted with Murat the fop; it is time that we forgot the eccentricities of his toilet, and considered the serious side of the King of Naples. King Joachim was a patient worker, and assiduous. He won and kept the affections of a woman as beautiful and remarkable as Caroline Bonaparte. He was a good husband and father, a self-sacrificing public servant. As much as this may perhaps be said of thousands of forgotten men. But the King of Naples was the originator, and the brilliant impersonator of great ideas. In the most difficult circumstances of his reign he bore himself with tact and honour. His kingly self-control in the face of outrageous impertinence was not the least remarkable of his many high qualities. When we contemplate the downfall of this king of men, it is a disagreeable element of the tragedy that fate put it into the hands of a second-rate man to overturn his throne, and that the man was an Englishman.

For the moment it seemed as if Lord William had toiled in vain. The downfall of the Emperor Napoleon did not entail the downfall of his brother-in-law. On the contrary, Joachim was more clearly king than ever. Freed from extortionate pressure from the side of France, he now found leisure for internal reforms; he approached them with sagacity and determination. Simple measures were enough to relieve the distress of Naples ; the reduction of taxes and the relaxation of the conscription were substantial reforms in the circumstances. It cost the king more, and was perhaps an even more important reform, to reduce the expenditure of his Court.

It is the attitude of England after the downfall of Napoleon that must be appealed to as evidence of the honesty of her intentions towards Sicily. The declared object of England had been to preserve the island for the House of Bourbon, keeping the Cabinet entirely clear of mainland politics. The plenipotentiary charged with the execution of these plans had chosen, however, to act in precisely the contrary sense. He had acted as if Sicily was destined to be English ground ; he had approached the Crown Prince on the subject of its transfer to the British Crown ; and he had resolutely refused to carry out the Austro-Anglian scheme for the pacification of the peninsula. He had openly stated that Sicily must remain

as a model self-governed state, the example of the proper form
of government to be adopted by Italy; and by words, and
ridiculous emblems he had pledged himself to an amount of
interference in the domestic affairs of the peninsula which he
might assert to be disinterested, but which nobody else could
believe to be anything but extravagant ambition. Contemporary observers could not be blamed for concluding that the
official instructions were a mere pretence. If they were
genuine, the attitude of England, after the abdication of
Napoleon, ought to be clearly defined; for according to
official declaration the necessity for the British occupation of
Sicily would disappear with the need of checking the emperor's
ambitious designs in the Mediterranean. It is agreeable to
be able to recite a series of acts which patently demonstrated
the honesty of England. The formal disavowal of Lord
William Bentinck has already been alluded to; but more
than that was necessary, if the cloud of suspicion which hung
over England's policy was to be dispelled.

The emoluments of the Sicilian Embassy were reduced
from £14,000 to £4000 a year. The latter amount was hardly
sufficient for the expenses of the post. It ranked Lord
William's successor with second-class Ambassadors, and
entirely took away the inflated importance of the Sicilian
Embassy. Moreover, the reduction was intimated in weighty
language to Mr. A'Court. He was informed that he was to
consider himself in future as an envoy merely, and not (as had
hitherto been the case) as the head of a Sicilian party. There
is much to be said against so sudden a change of attitude. It
exposed to persecution all those who had favoured the constitutional measures inspired by Lord William Bentinck. It
gave to King Ferdinand an opportunity (of which he
immediately availed himself) of taking revenge on Sicilians or
Neapolitans who had presumed to countenance the reduction
of his prerogatives and his revenues. All this gave a very
bad name to England, but it had the decided effect of dissoci-
ating the United Kingdom from the schemes of annexation with
which it had been credited. It might henceforth be said of

the English that they had been rash, thoughtless, indiscreet and careless of their friends ; but it could not be said with any show of reason that they had designed to annex Sicily. Still weightier evidence was to be forthcoming shortly in the attitude of England, among other European Powers, towards the question of Murat's title to be called King of Naples. On the side of Murat it was urged that his undecided attitude was the natural consequence of the hesitating policy of England— a policy notoriously in flat violation of that agreed upon between Prince Metternich and Lord Aberdeen. Murat was assuredly not to blame for this : and the treaty ought undoubtedly to be carried out at once, since the Allies had reaped the benefit of his neutrality. Against Murat all that could be urged was that, in point of fact, the treaty had not been ratified : and that as the Allies had no further need of his help, it need not be ratified. This was a plain proposal to take advantage of one's own *laches :* but shabby though this course was, it was, nevertheless, the only alternative for recognizing King Joachim forthwith and apologizing for the delay. The situation, already sufficiently complicated, was made still more difficult by the not unnatural distaste of the Royal families to welcome the House of Murat as one of themselves : and on the other hand the still greater difficulty of saying a good word for Ferdinand of Sicily. If the question had been settled, as was intended, by January 1814, it would already have been a thing of the past ; but it had not been settled, and there was a growing feeling in the Cabinets of Europe that it might be allowed to stand over till the Congress of Vienna. On this feeling the Duke of Orleans resolved to work. He made a long visit to France and England, and has left on record, in letters addressed to his father-in-law, an invaluable summary of the feelings of European sovereigns and statesmen towards the question of Sicily's position in the Mediterranean. Louis Philippe did not pretend to be an official personage. He treated the negotiations as a family matter. He carried no credentials, and merely approached the high personages with whom he dealt on the plain ground of trying to save his wife's father's property. This was natural ; and his ground having been wisely chosen he was received without any difficulty. To be received was one step ; but to effect anything after his reception was not so easy. He first approached the King of France, who was a relative, indeed ; but as the brother of Louis the Sixteenth he was not likely to feel any warm sympathy for the son of Egalité Orléans. He would do nothing for King Ferdinand, except refusing to acknowledge King Joachim. Nor would he consent to help the Duke of Orleans with letters of introduction in London ; but he allowed the duke to mention his name to the Prince Regent with the following equivocal message :—" Ask the Prince if he remembers that Knight of the Garter whom he received sitting." This might have brought Louis Philippe either a welcome or a rebuff : it was impossible to say what the king meant. But it was very clear that the King of France would not stir a single grenadier to turn King Joachim off his throne. The king was, no doubt, perfectly well advised too in keeping the soldiers of France well away from the sight of Murat's white plume.

From the Emperor of Austria, the Duke of Orleans received a cold but perfectly intelligible reply to his appeal in King Ferdinand's favour. The Court of Austria had behaved with great rectitude towards King Joachim. Deserted (as he could not help thinking) by England, the emperor had nevertheless persisted in the course to which he had pledged his word. There had recently been the exchange of decorations usual upon great occasions between Vienna and Naples. King Joachim had received the Grand Cross of the Austrian Order of Leopold. In the emperor's eyes, therefore, there was no Neapolitan Question. The ancient kingdom of the Two Sicilies had disappeared. Queen Caroline was described on her tomb in the vault of the Habsburgs as the wife of Ferdinand, King of Sicily. She had certainly been crowned queen of the Two Sicilies, and had so reigned for many years; but the Austrian Court was punctiliously careful of its obligations.

Not even in the privacy of sorrow was the new etiquette laid aside. Ferdinand was King of Sicily; Joachim was King o Naples. " Tell your father-in-law," said the emperor to Louis Philippe, " that he must give up all idea of returning to Naples. It is out of the question for him to think of it." The Emperor of Russia was even more decided, and added quaint comments and advice ; the comments very pointed, the advice hardly what might have been expected from the Tsar of all the Russias. "Tell your father-in-law," he said, "that peoples are no longer to be governed by holding out a hand to be kissed. Unless he can make up his mind to a really liberal and constitutional form of government, he must give up all idea of regaining the kingdom of Naples."

Thus the Duke of Orleans found no encouragement in Paris, and turned to London ; where, however, he found that the views of influential personages were even more opposed to his own than were those of the emperors of Austria and Russia. He could not induce the Prince Regent to admit that England had no engagements towards Murat. On the contrary, the conversation ran on the line that the only danger to Europe from a Bonaparte kingdom in Italy was the neighbourhood of the emperor in Elba. The Prince Regent was polite and even cordial ; but he could not be persuaded into regarding the question of Sicily as any other than one of numerous European questions to be solved according to expediency, and certainly not with any regard to the feelings of King Ferdinand.

" Your father-in-law has played his cards badly," was the Prince Regent's comment. The expression erred, perhaps, on the side of imputing to King Ferdinand some interest in the game, whereas he had, all his life, merely looked on. The Prince Regent remained to the end of the interview unmoved ; but furthered Louis Philippe's enterprise by presenting to him Lord Liverpool and Lord Castlereagh. The views of these two statesmen are even more significant than those of the reigning sovereigns. Lord Liverpool was quite determined. He laid it down at the commencement of the interview that England was indubitably bound to recognize King Joachim, and that in this matter Austria and Russia were with England. Louis Philippe was only too well aware that they were.

But, Lord Liverpool relentlessly went on, there remained France and Spain. There was nothing to hinder their joint action, except the obvious scandal, after all that had come and gone, of two Bourbon armies entering Italy to set up a third Bourbon throne.

But to suggest an offensive alliance between France and Spain, with the restoration of King Ferdinand as its objective, was to reduce the whole question to an absurdity. Louis Philippe was well aware that Louis the Eighteenth would never despatch an army to Italy on such an errand. With great readiness he changed his ground, and boldly taxed Lord Liverpool with being afraid of Murat. But in Lord Liverpool the Duke of Orleans had found his match; and the point of the attack was turned aside by Lord Liverpool laughing, and readily admitting the obvious fact that Murat would be a very awkward antagonist on the field of battle. In his turn he invited Louis Philippe to explain how he would propose to remove Murat from Naples: King Joachim's disappearance being the indispensable preliminary to King Ferdinand's restoration.

Louis Philippe was ready with his answer; he would confide the task, so he said, to Lord William Bentinck. Consciously or unconsciously he had touched the core of the matter; and in Lord Liverpool's rejoinder we have additional evidence of the attitude of the British Cabinet. Lord Liverpool became very grave at the mention of Lord William's name. Bentinck, he said, had been far too hasty, and had given Murat very just cause for offence.

This important interview throws much light on the events of the later years of the British occupation of Sicily. It appears from the words of Lord Liverpool just quoted, that there were no secret instructions issued to Lord William in a contrary sense to his published instructions. This being the case, the conclusion as to Lord William's action must be unfavourable, and it may be well to notice, before leaving the subject, what evidence has hitherto been left unconsidered, when we are estimating the events which preceded Lord William Bentinck's retirement from public life for thirteen years—until his appointment as Governor-General of India in the year 1827.

In the autumn of 1815 he proposed, uninvited, to winter at Naples. King Ferdinand had, by this time, returned to his capital on the mainland ; and on hearing of Lord William's intention he intimated to the British Ambassador that Lord William's presence was not desired by the Court of Naples. This is, in itself, remarkable ; since it was primarily and in great measure owing to Lord William Bentinck's refusal to recognize Murat that Ferdinand once more found himself in Naples as sovereign. What followed was still more remarkable. Bentinck would not have been Bentinck if he had paid any attention to hints. But the king was no longer dependent on England's favours, and he sent Lord William his passports. Even this hint was not strong enough, and Lord William continued, unmoved, his preparations to land and winter in Naples. Thereupon King Ferdinand stated plainly that if Lord William Bentinck dared to land, he would be arrested and turned out of the country by main force. There are incidents which defy comment, and this is one of them. But what was the attitude of England in the face of this affront to a lieutenant-general, a Knight of the Bath, a nobleman who had so recently held an exalted office ? Lord Liverpool insisted to Lord Castlereagh on "the importance of inducing Lord William Bentinck, if possible, to retire quietly out of Italy."

To return to Louis Philippe. Repulsed by Lord Liverpool, he turned to Prince Metternich, who told him plainly that he thought no more of King Ferdinand than the duke thought of King Joachim. Lord Castlereagh told him that everything must be left for the Congress. " I could make no way," he reported to his father-in-law.

But he would have been encouraged if he could have read

Lord Castlereagh's letters. It is clear that Castlereagh at this juncture, thoroughly disliking the position into which the military necessities of 1813 had forced him, was deliberately preparing to take advantage of the *laches* of England, and to restore King Ferdinand if possible. After the retirement of Lord William Bentinck from Sicily on July 14, 1814, it would have been no more than common honesty if King Joachim had been formally recognized forthwith. Delay, though not justifiable, was pardonable up to this date for the purpose of saving Lord William from mortification. It was not pardonable after July 14, 1814. Yet before this date Castlereagh was recording, with regret, that he could discover no grounds justifying us in changing our attitude towards Murat ; and in an interview with the Duke of Campochiaro at Vienna in October 1814, he developed the line of action which has to be recorded, but which is disagreeable reading for Englishmen.

Lord Castlereagh explained to the duke that it was unfortunate that Murat's hesitation had caused a delay in recognizing him as king before the Congress opened. The correspondence relating to the affairs of the Mediterranean must have been open to Lord Castlereagh. It is clear from that correspondence that Murat was as eager and accommodating as any party to a negotiation could possibly be, and that the delays were solely the work of the representative of England, who caused them by flat disobedience to the orders of his official superiors. It is also evident that the representative of England behaved towards a reigning sovereign, whom he was commanded to treat as such, with revolting insolence in speech, in action and in writing. Lord Castlereagh, however, instead of overwhelming the Duke of Campochiaro with apologies and regrets, went on to regret that the delay had prevented the question of "compensation" for Naples being brought more definitely before King Ferdinand. The language is inappropriate. It is not to a man of Ferdinand's character, in Ferdinand's position, that Great Powers are accustomed to extend "invitations." It was not by "inviting" that German unity or Italian unity was achieved. Austria, who was far more concerned to spare Ferdinand's feelings than England could be, understood this perfectly, when arranging with Murat to secure by force that Ferdinand should renounce his claim to the throne of Naples. Force was the only argument that King Ferdinand was capable of appreciating, as Lord Castlereagh must have been well aware.

Throughout the course of these negotiations, Lord Castlereagh suffered from the natural repugnance of a gentleman to the idea of repudiating a disadvantageous bargain. But he also suffered from the repugnance (which belonged to his country and the period in which he acted) to admit the sovereignty of any member of the Bonaparte family. His continual efforts to reconcile these divergent impulses produced the painfully undecided policy of England towards Italy during the years 1814 and 1815.

Thus on February 4, 1814, he wrote from Chatillon : " In proportion as Murat's support becomes less indispensable, one's repugnance to the arrangement in his favour has increased." A fortnight later he confessed to Lord William Bentinck " the British Government never liked this measure." But on March 30 (feeling, perhaps, somewhat ashamed at the treatment which the King of Naples was receiving at the hands of a British plenipotentiary) he issued the order that Lord William's " conduct towards Marshal Murat" should be regulated "upon principles of cordiality and confidence." Four days later he was called upon to repudiate formally Lord William Bentinck's extraordinary proposals to the Crown Prince in regard to the annexation of Sicily to England. At the same time he took the opportunity of pointing out that it was impossible to expect favourable results from the arrangements with Murat unless he was properly treated. As the bargain stood, Murat was to be recognized as King of Naples, while Ferdinand was to remain King of Sicily. In consideration of this concession on the part of Great Britain, the King of Naples was to employ his forces against the Emperor Napoleon. Bentinck was loud in his complaints against the lukewarmness of the Neapolitan sovereign. But was he not himself producing that lukewarmness by "a system of menace with, as he may suppose, the countenance of the British Government with respect to his title to Naples"? Such, in Lord Castlereagh's polite language, was undoubtedly Lord William's conduct—conduct which plainer people are accustomed to describe more severely.

There was only one course in honour open to England : to recall the Grand Duke of Würzburg to Tuscany, as had been proposed by Marshal Bellegarde in February ; to restore the other Italian sovereigns if necessary, and to shut up Murat within the lines of the ancient kingdom of Naples, with perhaps the addition of the Papal States. For the rest King Joachim should have been recognized at once ; there should have been no more of "Marshal Murat" in British despatches. The word of England had been given, and should have been redeemed. But Castlereagh allowed himself, somewhat against his will, to be swept away on the current of events. At the moment when he was interviewing the Duke of Campochiaro at Vienna on the subject of King Joachim's future, he had in his possession the Duke of Wellington's first sketch of a plan for the re-conquest of Naples. Sicily was to contribute ten thousand troops to this end ; Spain ten thousand ; France perhaps a few ; Portugal twelve thousand ; and England twelve or fifteen thousand. At the same time Mr. A'Court was desired to report on the "disposition of the nation to be placed under the old family."

All this is painful reading. But worse remains. "If Murat had but acted a decisive part," said Castlereagh to Campochiaro, "things might have been different ; but all his delays and vacillations have made his recognition impossible."

There is no man of honour who would not rather be vanquished than gain a victory by so shameful a distortion of the facts. The Duke of Wellington, although not asked for his opinion, except on the question of military detail, could not forbear a comment. "After all," he wrote to Lord Castlereagh on December 26, 1815, "our coming forward as principals is rather a delicate matter, under all the circumstances of the Austrian Treaty and the suspension of hostilities."

The Austrian Treaty and the suspension of hostilities were fast receding into the background as bases of objection to a Murat dynasty. It was the presence of the Emperor Napoleon in Elba that made the Powers anxious at the probability of a Bonaparte throne remaining established on the mainland of Italy. Not that this change of attitude was very important in the case of the Allies; for the establishment of Napoleon in Elba was even more definitely the work of the Allies than the alleged " delays and vacillations of Murat."

In the face of the collapse of every institution in which the English had taken pride as tending to ameliorate the condition of Sicily, some little modesty of language on the part of England would not have been out of place. If English statesmen had any justification for their authoritative language, it was to be looked for in the demonstrable superiority of their own methods of government to those adopted by the Sicilians. But after an indirect influence in Sicilian affairs lasting for more than twenty years, and an armed occupation which had lasted for five years, this was all that the British Ambassador could report as to what remained of English work in Sicily :—" The foundations of the constitution—or perhaps I should rather say, props, for foundations it never had any—are removed. "

At this juncture Lord Castlereagh was succeeded by the Duke of Wellington as British representative at the Congress of Vienna. Writing to the duke on the Neapolitan Question, Lord Liverpool urged the deposition of Murat, if it were found to be " just and practicable." " Practicable " was a question of military expediency for the Duke of Wellington to decide. As for the " justice " of expelling Murat from his kingdom of Naples, " this very difficult and delicate question," as Lord Liverpool described it, and as the Duke of Wellington described it, had been practically decided by both Lord Liverpool and Lord Castlereagh. Lord Castlereagh could have had no doubts as to where the " justice " of the question was to be found, for he had been compelled to rebuke Lord William Bentinck for his behaviour to King Joachim, and for his outrageous proposals to the Crown Prince of Sicily. Lord Liverpool, who had been driven to repudiate Lord William Bentinck, in conversation with the Duke of Orleans, was no more in the dark than Lord Castlereagh. There seems no reason, except desire to shift responsibility, for leaving the Duke of Wellington without definite instructions on the question. As a matter of fact, he was enjoined to remember that England would never consent to equip an armed expedition for the restoration of Ferdinand ; but beyond this negative instruction, he was desired merely to remember that Austria would probably not insist upon the retention of Naples for King Joachim, if the other Powers were opposed to the measure.

If the matter had been cognizable in a court of justice, King Joachim would have had a claim for specific performance which could hardly have been resisted. In consideration of certain acts to be done by him, England undertook to recognize him as King of Naples. The British agent prevented these acts being done, and then pleaded that King Joachim had avoided the performance of his duties under the agreement, and that England was, in consequence, absolved from her obligations : obligations which it had, by now, become distasteful to discharge. A more impudent course of action it would be difficult to imagine.

The Emperor Napoleon escaped from Elba ; and once more King Joachim renewed his offer of military support in exchange for the long-delayed recognition of his own sovereignty by England. Before entering on the final scene of the tragedy of Murat's death, the charges of "treachery" which abound in the correspondence of the time may be profitably considered. Negotiations undoubtedly went on between the King of Naples and the Viceroy of Italy. Seeing how recently the king had abandoned the cause of France, it was inevitable that the intimate relation, so long kept up between the two French marshals, now in opposite camps, should take some time in dissolving. If Murat gave encouragement to secret overtures, there is little to wonder at in his conduct, seeing the way in which England selected her agents and chose to conduct her negotiations. The "frankness," "honesty," and "straightforwardness" on which the British representatives were wont to congratulate themselves, are qualities more easily to be found in the manners of Murat and his Ministers. It is possible to be rude without being frank, and to bluster without being straightforward. Very finished examples of such behaviour are to be found in the transactions of England with the Neapolitan Court in the years 1813 to 1815.

Men who can so far lower themselves as to describe Murat's career as "a life of crime," or to write him down as "fool and knave," are not to be taken seriously. Such vulgar violence would discredit even a righteous cause. If Murat decided, by the spring of 1815, that nothing was to be expected from the word of England, he had only too good reasons for his conclusion.

He did so decide, and raised the standard of United Italy. He was forty-five years too soon. The people of Italy had a long and toilsome road to tread before they could rally round any one monarch. Besides, they were sick of warfare, and broken in spirit. Not even the unpopularity of Austrian rule could rouse them to resistance to the armies of Austria. For it is to be noted that the neutrality of Austria, rigidly observed, since the treaty of January 1814, was now at an end. Up to May 1815, King Joachim had no ground of complaint against Austria ; all his efforts had been directed to removing the hostility of England. But after fifteen months of anxious and fruitless negotiation he had decided that Austria was in league with England ; and that if he would preserve any shreds of sovereignty, he must act for himself. This was partly true. Metternich had always refused to advise his master to repudiate the treaty of January 1814. In that respect his attitude had been rigidly correct. But he counted on Murat being goaded into some act of indiscretion which could not be overlooked, and he counted rightly. By May 1815, the King of Naples had restrained himself as long as possible under what he described, with some justice, as persecution, betrayal and insult. His call to arms of the whole peninsula in the name of Italian unity, was regarded as a hostile move; and the armies of Austria were set in motion against him. He was defeated. On May 22, 1815, he fled from his capital and landed in the south of France. A fortnight later King Ferdinand returned to Naples. He had already consoled himself for the death of the archduchess ; and queen in fact, though not in name, Lucia Migliaccio took the place of Caroline of Habsburg. Joachim Murat made a wild attempt to recover his lost kingdom, and embarked for Calabria on September 28, 1815. There was no rising in his favour. He was seized immediately on landing, condemned to death, and shot on October 13, 1815.

" As an act of justice or an act of policy, his punishment is equally justified," was the comment of Mr. A'Court on this tragedy. " Justice ! " " Policy ! " We cannot always have both in measures of statesmanship, whether large or small. In English dealings with Italy during the years under review, it is to be feared that Italians may have good cause to complain that Italy could discover neither justice nor policy. Much that Italy suffered was inevitable : Italy itself was but a pawn in the complicated game which was being played out between France and England. But there is a residuum of injustice and impolicy for which England can hardly find an excuse. It speaks much for Italian patience and good-nature that Italians have allowed the memories of 1848, of 1860, and 1870, to obliterate the memory of 1790–1815.

INDEX

THE END